陳家廚坊
Chan's Kitchen

參乾寶貝

大師教路 烹調海味乾貨佳餚

Dried Seafood Cuisine

方曉嵐・陳紀臨
林長治 合著

萬里機構

前言

　　《參乾寶貝》是一本介紹海味乾貨的實用食譜書，「參」就是海參，「乾」就是乾貨，都是中餐食材中的寶貝。

　　海味乾貨的烹調比較複雜，由認識、鑑別，到浸發、入味等過程都需要一定的技術，尤其是海參，一般家庭主婦不容易掌握得很好。本書以海參為主角，並邀請了著名海味乾貨導師林長治先生參予提供資料；林先生對海味乾貨具有豐富的知識，特別是對海參的研究，他提供了有關海參的辨別和漲發的實用資料，加上我們對烹調海味乾貨的經驗，匯集成書，相信可以為讀者解決很多對烹調海參的疑惑，從而愛上烹調海參的菜式。

　　花膠魚肚，是近二十多年來城中最熱、女士至愛的乾貨。本書介紹多種常見的花膠魚肚，方便讀者去識別及選購；不同的品種，以不同的方法去浸發和應用，以美食補身之餘，更增添入廚樂趣。

　　江瑤柱是我們自小認識的乾貨，它是粵菜中經常採用的食材，幾乎家家戶戶或多或少都有採用，是海味乾貨中最為常見。

其他乾貨如魷魚、墨魚、蝦乾，以及比較少見的瀨尿蝦乾，都為你一一道來，令菜餚更加精彩。

本書詳細闡述了海參和花膠魚肚的品種、營養價值、如何選購、儲存，以及不同乾貨的浸發方法，並為讀者介紹了數十個易做的相關菜式，其中有請客吃飯的雅宴菜式，也有平常的家庭菜，是一本集海味知識和食譜書功能的實用參考書，絕對值得珍藏。

本書得到著名的法國麗固瓷器公司借出精美餐具，使我們的菜餚顯得明亮生輝，倍增貴氣，特此鳴謝國際著名餐具瓷器設計師張聰先生。

同時，再次感謝海味乾貨老行尊林長治先生的食材贊助和提供寶貴的知識。

方曉嵐

2021 年夏

目錄

認識
海參乾貨

不同產地的刺參有甚麼分別？怎樣浸發不同品種的海參，如禿參、刺參、豬婆參？怎樣做海參才入味？

女士們至愛的養顏補品花膠，究竟和魚肚有甚麼區別？怎樣浸發不同種類的花膠和魚肚，例如：白花膠、鱉肚、紫膠肚、鱔肚？

無論做主料或配料都非常出色的江瑤柱、蝦米、瀨尿蝦乾、土魷、墨魚乾和鱘魚，要怎樣炮製才能更特顯它的鮮、香、濃郁？

我們都會為你一一道來，並配上相關的菜餚；期盼我們的「參乾寶貝」也成為你們的心肝寶貝。

海參

海參早在 6 億多年前的前寒武紀就已經存在,但兩
千年前才有正式文獻記載。牠是一種名貴海產動物,
屬於高蛋白質、低膽固醇,而且脂肪非常少,幾乎接
近零的食品,一直都被當成天然養生珍品,故有「大
海之珍」的美譽,亦是「中國八珍」之一。由於廣泛
生長在世界各大水域,故不同地域的名字也有不同,
英文是 sea cucumber,法文是 bêche-de-mer,馬
來西亞文是 trepang,日文是 namako 和菲律賓文是
balatan。

海參的生長條件

　　海參屬於棘皮動物門海參綱的無脊椎動物,亦稱作「海
黃瓜」或「海鼠」,主要棲息於熱帶珊瑚礁區、礁岩岸、泥
沙海底或隱藏在石塊下,結聚一起,加上不善運動,利用
管足的吸盤匍匐而行及吸附岩礁,或是借用體壁肌肉收縮
行動。海參的身體以長筒狀或圓柱狀為主,體表顏色大多
呈黑色、褐色、灰白色等,隨着棲息環境而變化體表,利
用保護色潛伏沙地或礁石生活。怕熱和超強的再生能力是
海參的特質,口在前端,周圍生有許多伸縮自如的觸手,
並以觸手的分泌黏液獵取四周食物顆粒如藻類、有機碎
屑,或腐爛食物等,然後送入口中餵食,其消化道兩側相
對而成,稱作呼吸樹的呼吸器官,當海水經由肛門進入體
內,繼而在呼吸管內與體腔液間行呼吸作用,可説是用「臀
部」呼吸。同時,海參的體壁和觸手等也會進行呼吸作用。
牠的肛門在後端,剛好與口部為兩極而生,由於海參與隱

魚互為共生的關係，當隱魚遇到敵害時，就飛快地鑽入海參的肛門中避難。

　　海參扮演着海洋生態系統的海底清道夫，幾乎沒有天敵，只是遇到人類採捕或其他魚類的誤食時，因受到強烈刺激、過度擠壓或缺氧的情況下，自割體壁分成兩段逃敵。有些海參會利用體壁的劇烈收縮，從肛門內射出居維氏管束，它猶如呈粉紅色或橘色帶黏性的細絲線一樣，因該黏絲線含有異味〔即「海參毒素」（holothurin）〕，可干擾獵食者糾纏。

　　海參是一種活動緩慢的海洋生物，習性很容易掌握，攝食有規律，最適合生活於 2-40 米深的海底，水溫宜在 0℃-28℃，至於海水鹽度為 28-31%，其繁殖期在每年的 6-7 月，有趣的是，牠們可進行無性分裂生殖法來傳宗接代，故只需把身體自行分裂解體，便可繁殖生長，而其傷口會藉助超強再生能力而在短時間封合。

　　海參分佈在溫帶和熱帶的海洋，水流緩慢的沙地、岩礁和珊瑚礁，為海參提供了豐富的食物，例如印度洋、西太平洋、西沙群島、日本北海道、台灣、遼寧和山東沿岸等地區均是海參的理想棲息地。

海參的營養和食療價值

　　海參具有多種營養成分，包含蛋白質、脂肪、碳水化合物、粗纖維、維生素 A、鈣、磷、鐵、碘等。牠同時有少量三磷酸腺甙（ATP），較高的光氨酸、精氨酸、松氨酸等。牠的蛋白質為水溶性，不需要鹽、酸、鹼及脂肪的幫助即可分解為各種極易被人體吸收的氨基酸。再者，其成分中有 18 種氨基酸，當中有 10 種均是人體高需求的，特別是構成男性精細胞的重要成分的精氨酸，含量遠遠高於其他海產品，還有合成人體膠原蛋白，可促進細胞再生和機體損傷後修復，以及提高淋巴細胞的免疫活性，增強人體免疫力，延年益壽。

海參的成分可與人參媲美，特別是含有海參毒素；牠的結構很像皂甙（saponin）的配醣體化合物，又稱「海參皂甙」，與「人參皂甙」相似。此皂甙在海參的體壁、內臟和居維氏管束內找到，當中又以居維氏管束的含量最多，也含酸性黏多糖、膠原蛋白。以上這些特質才是海參贏得「食中藥王」的美譽由來。

海參的品種擇要

全世界大概有 900 多種海參，可供食用的只有 40 多種（中國約有 101 種，供食用的僅有 21 種）。根據海參背面是否有圓錐肉刺狀的疣足分為「有刺參」和「光皮參」兩大類。其中「刺參」主要有「刺參科」的種類，「光皮參」主要有「海參科」、「瓜參科」和「芋參科」等種類。一般會採用「海參科」、「刺參科」、「瓜參科」和「芋參科」等加工食用。

海味店裏常見的海參

香港海味店常見的海參品種，以刺參、石參、禿參為主，這些海參主要來自中國、日本、俄羅斯、澳洲、印尼、菲律賓、中東、美國、韓國等地，特別是刺參或遼參，主要來自日本關東和關西以及中國山東、遼寧和俄羅斯的符拉迪沃斯托克。石參的貨源則以美洲、印尼、澳洲和中東為主，大型的豬婆參多來自印尼、澳洲、非洲及所羅門群島，分有黑參和白參，肉厚肥胖，帶有砂泥或灰質，處理時需要用火燒燶以去灰泥；但一些烏石參則已沒有灰砂，可刪減火燒的步驟。一般家庭自用可選用來自西非、中東、墨西哥和澳洲的禿參，處理比較方便及容易。

常吃海參的品種

✦ 刺參 / 遼參 ✦

刺參生活於海流平靜的海灣內，通常棲息於 3-15 米的岩礁或細泥沙海底，體長一般在 6-20 厘米之間，來自山東、遼東半島、俄羅斯海參崴、日本、南北韓和南美洲等地的水域。這種海參的身體背部佈滿大小不等的肉刺，故稱為「刺參」；因為刺參盛產於中國山東半島和遼東半島，品質上乘，食用歷史悠久，故亦稱為「遼參」。

乾刺參體呈圓棍狀，兩端鈍圓，腹面較平坦，背部開膛，上有 4-6 行縱刺，淺黑或淺灰色。參體肥壯、飽滿、順挺，肌肉厚實，肉刺挺拔鼓壯，體表無殘缺下陷，刀口處向緊厚外翻者為上品。相反地，參體枯瘦肉薄、坑陷大、歪曲不挺直，肉刺倒伏、尖而不直，圓椎偏小，體表有潰爛殘跡者為次品。

香港出售的刺參乾品以日本的品質最好，味香爽脆，但價錢也較高。刺參體色除俄羅斯、日本產地有黑色和赤色，而中國渤海灣的加工參體，正宗為灰色。日本北海道的關東刺參的肉刺較幼而長，但較關西刺參為密；日本中部出產的關西刺參肉刺較疏而短。

日本關東刺參（又稱北海道刺參）

體型肥壯、肉厚；因關東水域略冷，肉刺挺拔肉針略尖而長；色澤黑色光亮；體表無殘跡缺刻，刀口處的肉緊厚外翻。

成品製作嚴謹，講求標準，乾度十足。

日本關東刺參

日本關西刺參

外形貌如其名，身上長滿肉刺，體積最長者為 4-5 吋，一般 3-4 吋或 2-3 吋居多。

因關西水域暖，日本關西刺參肉身略平短。

參體流線形不及關東刺參的外形優美。

日本關西刺參

海參

日本沖子刺參

　　日本沖子刺參屬於日本沖繩較新的海參品種，售價比
關東刺參或關西刺參便宜，是一種入門級的刺參。

日本沖子刺參

大連刺參

　　中國山東半島和膠東半島的威海海域因地形特別，能伸入海中最深處，三面環海，黃海渤海交匯處，水流較快，無污染、海藻品種豐富，氣候獨特，水溫合適，當地出產的刺參，肉刺硬、挺、長，呈錐形，品質高。其次數煙台長島、蓬萊刺參，地處渤海邊，水質和氣候較為適宜，刺參品質較佳，接着就是大連遼參。

　　海參崴的刺參和大連、山東的遼參外形極其相似，海參身上的刺針較長，很似「狼牙棒」。

野生大連刺參

✦ 南美刺參 ✦

　　產自南美洲的刺參，由於生長在水溫較暖的海域，身形粗大肉厚，刺針圓短，類似像「流星鎚」般。灰味較重，其爽脆程度和韌性不及生於冰凍海域的同類型海參，但勝在圓潤肉厚。其出產海域大多在中南美洲，如墨西哥灣、巴西、哥倫比亞、千里達等。

　　刺參按品質排位，第一位當屬產自日本、中國、海參崴的刺參；第二位才到南韓、北韓、南美洲等地的出品。

南美刺參（產自墨西哥灣）

✦ 白石參 ✦

白石參的名稱乃直譯自英語 White stone，又名「豬婆參」；與刺參相比，白石參肉多而軟滑，刺參則較爽，可說各擅勝場。這類海參的產地來自印尼、菲律賓、所羅門群島、南北也門、斐濟群島，中國的南海中沙群島也有出產。尤以印尼所出產的較為著名和上等，俗稱「鬼仔摔」，屬軟皮海參，雖經晾曬後，參體仍保留柔軟的特質，可任意扭擰。

與牠同類的還有烏石參、港石參、雁石參、非洲石參、雪石參等。港石參皮帶幼嫩，雁石參則皮殼略硬，此兩種石參每 600 克可發多至 3-4 公斤。烏石參肉厚而脸滑，非洲石參和雪石參每 600 克只發 1-2 公斤。

乾貨表面光滑無刺，色白帶黃，水發前必須用火炙燎其皮，否則不能脹發。

石參一定要挑選外形肥大飽滿，乾度十足，不黏手的，這才是好貨。反轉參體，可以稍為彎曲的更是上等好貨。

支數代表參體的斤兩，如 1 斤有 2 支、3-4 支或 4-6 支等，一般情況下，支數愈少，表示參體發脹愈大。最好的海參當然是 1 支為 1 斤重，但貨源極少，1 斤 2 支較普遍。

海參

白石參

生長於淺水域，咖啡色，石灰重，皮層略厚，肉身厚4-6厘米。源自東南亞、中東、所羅門、也門群島等水域。

體呈筒狀，略扁，體表光滑，背部有肉粒突出，腹部有粉白狀石灰質。

皮薄肉厚，觸手突出似豬乳頭，因此香港海味供應商稱之為豬婆參，沿用百年。

印尼白石參

黑石參

　　黑石參的名稱是從英語 Black stone 直譯而來，又名烏參。牠屬大型海參，黑色，體重可達 8 公斤，及後經晾曬乾燥製成的乾海參，外皮天然烏黑且帶硬，但內裏肉色卻是色白肉厚。由於肉厚而腍滑，故頗受歡迎。其產地來自所羅門群島、印尼、汶萊等國家。

　　這種海參屬家庭常用參，以乾身、堅挺硬實，色澤鮮明為要。市面一些海參專門店或海味店有代客浸發服務，亦有已浸發的急凍貨。

　　黑石參皮層比豬婆參薄，石灰質較少，一般體型，闊 5 厘米，長 25 厘米。

黑石參

❖ 禿參 ❖

　　禿參的名稱，鑑於牠從海上撈上岸時就沒有皮殼，光禿禿的，全條皮黑，肉白光亮，故而名之。

　　其體型不大，漁民在加工時往往不剖肚洗沙就用水煮，也不拌石灰便曬製，所以往往參體內裏藏有大量沙粒；若在浸發時沒能處理乾淨，吃起來就會是滿口沙子了。雖然如此，其皮薄肉厚，亦是受歡迎的貨色。

　　牠的產地是所羅門群島、關島、印尼、非洲，以馬達加斯加品質最佳，雪白通透，而澳洲出產的品質亦佳。

馬達加斯加禿參

澳洲禿參

斯里蘭卡白玉沙禿參

南美禿參

如何選購海參

消費者要挑選一條好海參，可以利用觀看和手觸判斷，可依從以下基本守則：

身體完整，體表光澤，肉壁肥厚，個頭齊整均勻，乾燥度十足，開口端正；刺參肉刺堅挺、石參和禿參要體態齊整無缺。

海參的浸發方法

浸發海參，一靠方法，二靠手感，三靠經驗。

海參浸發的方法很簡單，以不同的種類，分為三種基本的浸發方法；二靠手感，就是在浸發差不多完成時，以手去輕輕觸捏海參，測試軟硬度。海參是生物，質感不可能隻隻相同，每批貨的品質也可能不相同。另外，對海參的口感軟硬，更是每個人的喜好不同，覺得不夠軟，就再焗水浸一次。第三當然是靠經驗，做得多了，自然會掌握得更好，更合自己的心意。

浸發海參可以用玻璃鍋或合金鍋，但應該避開無塗層的生鐵鍋，以免受到氧化的影響。坊間流傳一個說法，就是不能用有油的鍋。我們曾經做過實驗，分別用玻璃鍋和加了豬油的合金鍋浸發遼參，結果是沒有任何分別，發好的遼參的重量是乾遼參的 6 倍，肉質也沒有分別。所以我們認為發海參只要用乾淨的鍋就可以了。

✦ 刺參的浸發 ✦

　　一般情況，乾品長 7.5-12.5 厘米的刺參、遼參或南美刺參，浸發約需 2 至 3 天時間。5 厘米以下的乾刺參，只需隔晚準備。細支刺參幼嫩，只需煮 5 分鐘便可熄火原鍋焗浸，反覆滾、焗、浸兩次即可。

浸發過程如下：

1. 首天先燒清水至沸騰，放下乾刺參，煲 15 分鐘，熄火焗 8 至 12 小時。此程式在換水後再重複做一次。
2. 兩次煮焗浸水之後，海參發脹了幾倍，用刀在底部直線剐一刀，使參體內的腸臟吸水變脹大。接着，燒清水至沸騰，放下已脹發的刺參煮 10-15 分鐘，熄火原鍋焗 8 小時，撈起，取出內臟，並把用小刀刮掉內壁（衣），避免微幼沙粒殘留參體，浸發程式便告完成。

　　以下是刺參浸發前後的參體變化：

已浸發：
10 厘米

乾品：
5.5 厘米

日本關東刺參

乾品：
5.5 厘米

已浸發：
12 厘米

日本關西刺參

乾品：
5 厘米

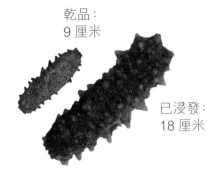

已浸發：
11 厘米

野生大連刺參

乾品：
9 厘米

已浸發：
18 厘米

南美刺參（產自墨西哥灣）

乾品：
8.5 厘米

已浸發：
17 厘米

日本沖子刺參

禿參的浸發

禿參浸發過程通常要 3-4 天，大隻的甚至 5-6 天才會鬆軟，方法如下：

1. 先煲大滾水，水滾放下大約 10-20 厘米的禿參，滾 20-30 分鐘，熄火焗 3 小時撈起浸清水，放在雪櫃浸 24 小時，期間最少要換水一次。
2. 翌日，海參發大脹身，這時從海參中間直線剢開一刀，再分開稍硬和軟的海參。再煲大滾水，先放下手感較硬的海參，滾煮 15 分鐘，再放入較軟的海參，一起煮 10 分鐘，熄火焗 1-2 小時後，撈起浸凍水 12 小時，期間換水一次。
3. 經過兩天的滾、焗、浸，海參內臟可以挖出來。但若有部分仍然實心未發脹透，可以再用滾水浸、焗 15 分鐘至 1 小時，撈起浸清水 8-10 小時，便可以取出腸臟。

注意：體型較細小的禿參，例如來自斯里蘭卡的白玉沙禿參，可以用刺參的浸發方法。

以下是禿參浸發前後的參體變化：

乾品：
5 厘米
已浸發：
16 厘米

斯里蘭卡白玉沙禿參

乾品：
長 12 厘米、直徑 2.5 厘米

已浸發：
長 20 厘米、直徑 6 厘米

馬達加斯加禿參

乾品：
10 厘米

南美禿參

已浸發：
19 厘米

乾品：
8 厘米

已浸發：
16 厘米

澳洲禿參

石參（豬婆參）的浸發

石參的浸發較複雜，因為它體大肉厚潺多，牠的乾製過程會用大量石灰處理，故乾品表面仍留有石灰粉，難以清除。酒家採用傳統的方法是用火燒其表面，將石灰連皮殼燒去後再浸清水，但這方法在家庭中不宜採用；因為在燒灰皮期間，灰塵飛揚，濃煙怪味滿屋，說不定鄰居會誤以為發生火警。以下提供兩種比較簡單的方法，讓大家在家也可以自己發石參。

方法一：

1. 燒水至沸騰，放入乾石參煲 30 分鐘，熄火後原鍋焗 3 小時，再取出放清水浸泡一天，期間要換水 2 次。第二天同一程序重複一次。

2. 第三天，換清水，燒至沸騰，放入已脹發的石參煲 30 分鐘，熄火後原鍋焗 2 小時，然後換清水放入雪櫃浸至皮殼脹大，用百潔布大力擦去表面灰分，如遇凹凸坑位未能擦去，再重複焗水一次。

3. 經過四、五天重複浸焗，此時石參已脹大，可用尖刀劏開，取出內臟和沙粒，浸發過程完成。

方法二：

1. 先煲大滾水，水滾放下大支裝石參，用文火煲 30 分鐘，熄火焗 2-3 小時。

2. 撈起海參，浸清水 10-15 小時，第二天同一程序重複一至兩次。經過兩、三次的滾、焗、浸等過程，參體會脹大很多。

3. 以薑煲水，燒滾後再放下已浸軟的海參，加入 1 杯玉冰燒酒，用文火煲 30 分鐘，然後浸清水放入雪櫃 48 小時，才可清理表層沙粒和灰分，劏開取去肚內腸臟，便完成浸發過程。

以下是石參浸發前後的參體變化：

乾品：
9 厘米

已浸發：
17 厘米

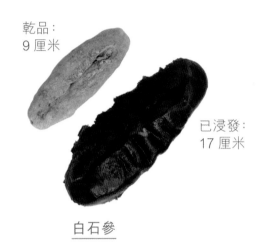

白石參

乾品：
10 厘米

已浸發：
18 厘米

黑石參

海參入味的方法

　　海參的營養價值很高，是很有益的食材，但海參本身沒有海產的鮮甜味，可以說是完全無味。用海參入饌，除了用來煲湯，或者加以濃味汁醬的涼拌、燜燒或爆炒，否則就需要預先入味。

　　發好的海參，入味的方法是用上湯或濃雞湯，有蒸和煮兩個方法：

1. 要保持原狀的名貴海參，例如遼參或日本刺參，或者已經發得過分柔軟的海參，不建議放入湯中滾煮，而是浸着湯來蒸。因為用煮的方法，就算用很微的火，也有可能把海參煮爛。用蒸法的原因是，浸着湯的海參是不動的，比較容易保持完整。豬婆參因為比較大，也建議用蒸的方法入味。

   撇去上湯或雞湯表面的油分，把海參浸過面，大火蒸 20 至 30 分鐘，海參即可入味，蒸的時間長短，按海參的大小厚薄來掌握。

2. 一般的海參，例如禿參，或不容易入味的白玉參，可放在已撇去表面油分的上湯或雞湯中，煮滾後用小火熬煮 10 至 15 分鐘至入味。煮的時間長短，按海參的大小厚薄來掌握。

重要注意事項：

　　無論採用煮或蒸的方法，用來使海參入味的上湯或雞湯，必須是無鹽或盡量無鹽的湯，浸發好的海參遇鹽加上長時間加熱，就會慢慢縮細，失去浸發好的形狀。由於現成的雞湯，已經含有鹽分，建議用土雞或老母雞熬成濃雞湯，切記不要放鹽及火腿，因火腿含很高鹽分。

海參

涼拌魚香鮑魚海參

Abalone and Sea Cucumber Salad
with Fish Flavored Sauce

份量：4人份 ｜ 準備時間：15分鐘 ｜ 烹調時間：5分鐘

建議採用

馬達加斯加禿參

斯里蘭卡白玉沙禿參

南美禿參

澳洲禿參

調味醬

材料

麻油：1茶匙

胡椒粉：1/2茶匙

鹽：1/4茶匙

頭抽：1/2茶匙

黑醋：1湯匙

糖：1湯匙

紹興酒：1湯匙

材料

葱：2條，切絲

郫縣豆瓣醬：1/2湯匙，剁碎

蒜蓉：1湯匙

薑蓉：1湯匙

罐頭鮑魚：2至3隻

發好海參：300至400克

做法

1. 發好的海參汆水1分鐘，撈出瀝乾，切成5厘米長1厘米粗的長條。

2. 鮑魚切片，再切成和海參相約的長度。

3. 燒熱2湯匙油，爆香薑蓉、蒜蓉和豆瓣醬，加入調味醬材料煮成魚香醬，放在碟底，上面放海參、鮑魚和葱絲，吃時拌勻。

海參

英文食譜見第153頁

蝦球海參燒椒汁

Shrimp and Sea Cucumber Salad
with Chili Pepper Dressing

份量：4人份　準備時間：15分鐘　烹調時間：15分鐘

建議採用

馬達加斯加禿參

斯里蘭卡白玉沙禿參

南美禿參

澳洲禿參

材料

發好海參：300克

中蝦：300克

湖南辣椒：4隻

檸檬汁：1湯匙

鹽：1/4茶匙

糖：1/2湯匙

黑胡椒碎：1/2茶匙

橄欖油：1湯匙

做法

1. 海參汆水，放涼後切成2厘米方塊。
2. 蝦去頭剝殼，挑出蝦腸，汆水至熟，瀝乾放涼。
3. 去掉湖南辣椒的籽，放在火上烤至表皮半焦，連皮剁碎。
4. 把辣椒碎、檸檬汁，黑胡椒碎、鹽、糖和橄欖油拌勻成燒椒汁，再放入海參和蝦即成。

海參

英文食譜見第154頁

燒椒汁的靈魂是湖南辣椒和檸檬汁，味道清新。

沙茶海參

Sautéed Sea Cucumber
with Satay Sauce

份量：4人份 ｜ 準備時間：15分鐘 ｜ 烹調時間：3分鐘

建議採用

馬達加斯加
禿參

斯里蘭卡白玉
沙禿參

南美禿參

澳洲禿參

材料

發好海參：300 克

芥蘭：150 克

蒜頭：2 瓣，切片

沙茶醬：1 湯匙

糖：1/2 茶匙

紹酒：1 湯匙

生粉：1/2 茶匙

做法

1. 發好海參汆水，瀝乾，切成約 5 厘米長 1 厘米粗的長條。

2. 芥蘭洗淨，切 5 厘米段，用水焯至僅熟。

3. 燒熱 2 湯匙油，爆香蒜片，加入沙茶醬、糖和海參大火爆炒，潷酒，放入芥蘭同炒，用生粉勾薄芡，即成。

海參

英文食譜見第 154 頁

宮保蝦球海參

Prawns and Sea Cucumber with Gongbao Sauce

份量：4 人份 ｜ 準備時間：20 分鐘 ｜ 烹調時間：15 分鐘

建議採用

馬達加斯加禿參

斯里蘭卡白玉沙禿參

南美禿參

澳洲禿參

調味醬

紅油：1茶匙	麻油：1茶匙	鹽：1/4茶匙	生粉：1/2茶匙	醋：1/2湯匙	頭抽：1/2茶匙	胡椒粉：1/4茶匙

材料

花生：30克	葱花：2湯匙	蒜片：10克	薑片：5克	花椒：1茶匙	青紅甜椒：各半個，切塊	小紅椒1至2隻，切段	生粉：2湯匙	中蝦：8隻	發好海參：250克

做法

1. 海參汆水，瀝乾，切成 1.5 厘米粒。
2. 蝦剝殼，從背部剖開，去掉蝦腸，洗淨，瀝乾。
3. 花生炸好，瀝油。把調味醬材料在碗中拌好，用前再拌勻。
5. 把海參和生粉拌勻，炸脆，撈出瀝油。
6. 燒熱 1 湯匙油，爆香小紅椒、青紅甜椒、花椒、薑、蒜和葱，放入海參和蝦球，加入調味醬炒至蝦熟及汁稠，拌入炸好的花生，即成。

海參

英文食譜見第 155 頁

葱燒海參

Braised Sea Cucumbers
with Beijing Scallions

份量：4 人份 │ 準備時間：10 分鐘 │ 烹調時間：30 分鐘

建議採用

南美刺參

大連刺參

日本沖子
刺參

北海道
刺參

材料

麻油：1 茶匙

生粉：1/2 湯匙

胡椒粉：1/8 茶匙

糖：1 茶匙

蠔油：2 湯匙

大葱（京葱）：2 條

薑汁：1 湯匙

無鹽雞湯：125 毫升

發好刺參：4 條

做法

1. 把發好的海參洗淨；在雞湯中加薑汁，浸入海參蒸 15 分鐘。

2. 大葱葱白切成 5 厘米長段，一半留起，把另一半葱段（做葱油），
   破開兩邊，用 6 湯匙油小火煎炸至出味，葱段不要，葱油留用。

3. 燒熱 3 湯匙葱油，先煎香葱段至金黃，取出葱段。在葱油中放
   入海參和雞湯，煮沸。

4. 放入蠔油、糖、胡椒粉，燒煮約 5 分鐘，放入葱段，中火收汁，
   再用生粉勾芡，加麻油兜勻，盛碟，再淋上 1 湯匙葱油即成。

海參

英文食譜見第 156 頁

瑤柱花膠伴海參

Dried Scallop, Sea Cucumber and Fish Maw

份量：1 人份 | 準備時間：30 分鐘 | 烹調時間：45 分鐘

建議採用

大連刺參

北海道
刺參

材料

麻油：1/2 茶匙

生粉：1/2 湯匙

無鹽雞湯：250 毫升

蠔油：1 茶匙

紹興酒：1 湯匙

薑汁：1 茶匙

發好花膠（紮膠公）：1 件

發好刺參：1 條

宗谷江瑤柱：1 粒

做法

1. 江瑤柱用清水浸過面泡 30 分鐘，連水蒸 30 分鐘，取出。
2. 把江瑤柱、花膠和刺參排好在深碟，加入雞湯、薑汁和酒，蒸約 15 分鐘，小心倒出雞湯汁。
3. 把雞湯汁加蠔油，用生粉勾薄芡，拌入麻油，淋在材料上即成。

海參

英文食譜見第 157 頁

蹄花海參

Braised Sea Cucumber
with Pig Trotters

份量：4 人份 | 準備時間：15 分鐘 | 烹調時間：2 小時

建議採用

馬達加斯加
禿參

澳洲禿參

斯里蘭卡白玉
沙禿參

南美禿參

材料

豬蹄：500 克

發好海參：300 克

薑片：50 克

葱：4 條，切段

紹興酒：2 湯匙

無鹽雞湯：250 毫升

頭抽：1.5 湯匙

糖：1 湯匙

鹽：1/2 茶匙

葱油：1 湯匙

做法

1. 豬蹄汆水，再煲 1.5 小時，沖冷水，脫骨，切成 2.5 厘米大小方塊。

2. 海參洗淨，汆水，瀝乾，切成如豬蹄大小的塊。

3. 在鑊中下 1 湯匙油，爆香薑和葱，放入豬蹄和海參，瓚酒，加
入雞湯，大火煮沸，轉小火，放入頭抽、糖和鹽，煨煮約 15 分
鐘至入味。

4. 大火收汁，加 1 湯匙葱油拌勻，即可盛碟。

海參

英文食譜見第 158 頁

菠蘿咕嚕脆海參

Sweet and Sour Sea Cucumber

份量：4人份 ｜ 準備時間：5分鐘 ｜ 烹調時間：25分鐘

建議採用

斯里蘭卡白玉沙禿參　馬達加斯加禿參　南美禿參　澳洲禿參　南美刺參

材料

發好海參：300克
無鹽雞湯：125毫升
生粉：3湯匙
罐頭菠蘿：1小罐
紅甜椒：1/2個
青甜椒：1/2個
洋葱：1/2個
蒜頭：2瓣，切片
大紅浙醋：4湯匙
紅糖：4湯匙
麻油：1茶匙

做法

1. 發好的海參切段，用雞湯煮10分鐘入味，撈出，拌入乾生粉，炸脆。
2. 甜椒去籽切塊，菠蘿和洋葱切塊。
3. 浙醋和紅糖拌勻成糖醋。
4. 燒熱2湯匙油，爆香蒜片和洋葱，加入糖醋煮稠，再放入海參、菠蘿和甜椒，爆炒約1分鐘，用少許生粉勾薄芡，拌入麻油即成。

海參

英文食譜見第159頁

肉片炒海參

Sautéed Sea Cucumber with Pork

份量：4人份 | 準備時間：10分鐘 | 烹調時間：3分鐘

建議採用

馬達加斯加
禿參

斯里蘭卡白玉
沙禿參

材料

發好海參：300克

醃頭豬肉：125克

韭菜花：125克

鹽：1/4茶匙

糖：1/4茶匙

生粉：1/2茶匙

蒜頭：2瓣，切片

蠔油：1湯匙

生粉（勾芡用）：1/2茶匙

麻油：1茶匙

做法

1. 豬肉切片，用鹽、糖和2湯匙清水醃10分鐘，再加生粉拌勻。

2. 海參斜切厚片，汆水1分鐘，瀝乾。

3. 韭菜花洗淨，切5厘米段。

4. 燒熱2湯匙油，爆香蒜片，放入豬肉和海參爆炒約1分鐘，加入蠔油和50毫升水，煮沸，放入韭菜花炒熟，用生粉勾芡，加麻油即成。

海參

英文食譜見第160頁

百花釀遼參配蔥汁

Stuffed Sea Cucumbers

份量：4人份 | 準備時間：30分鐘 | 烹調時間：15分鐘

建議採用

大連刺參

北海道
刺參

材料

發好刺參：4 條

無鹽雞湯：150 毫升

薑汁：1 湯匙

鮮蝦：300 克

粗鹽：1/2 茶匙

肥豬肉：10 克，切細粒

雞蛋白：1/2 個

幼鹽：1/8 茶匙

糖：1/8 茶匙

白胡椒粉：1/8 茶匙

生粉：1/4 茶匙

蔥汁

蔥：6 棵

日本芥末：1 茶匙

鹽：1/8 茶匙

海參

英文食譜見第 161 頁

1. 把發好的海參洗淨，在雞湯中加入薑汁，浸入海參蒸 30 分鐘，雞湯留用。

2. 鮮蝦去頭殼，挑出蝦腸，放在笪箕裏，加粗鹽用手抓洗，再用冷水沖淨瀝乾；把砧板洗淨抹乾，把蝦仁用菜刀刀身壓扁，然後再用刀背反覆粗剁成蝦茸，用刀把蝦茸鏟起放在大碗中。

3. 在蝦茸中加入蛋白，用筷子循同一方向攪拌至起膠，再加入幼鹽、糖、白胡椒粉和生粉等拌勻，用手抓起蝦茸再撻回碗中，如此反覆多次，至撻至呈膠狀。

4. 在蝦膠中加入肥豬肉粒，再釀入海參中，大火蒸 10 分鐘。

5. 葱只用葱青部分，汆水 10 秒，泡冰水至涼，切段，和雞湯、芥末和鹽一同攪拌成葱汁，與釀海參同上，即成。

蝦籽大烏參拌葱黃

Braised Sea Cucumber with Shrimp Roe and Beijing Scallions

份量：4 人份 | 準備時間：1 小時 | 烹調時間：15 分鐘

建議採用

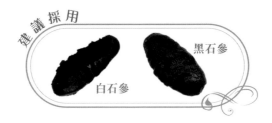

黑石參

白石參

材料

發好大烏參：1 條

無鹽雞湯：250 毫升

蝦籽：2 湯匙

大葱（京葱）：4 條

薑片：30 克

蒜頭：2 瓣，切片

紹興酒：2 湯匙

糖：1 湯匙

鹽：1/2 茶匙

老抽：1 湯匙

生粉：適量

海參

英文食譜見第 162 頁

做法

1. 用無鹽雞湯把發好的海參浸過面，大火蒸 30 分鐘。

2. 用白鑊微火把蝦籽炒香，盛起備用。

3. 大葱白切成 4 厘米長的段，一半留起，把另一半的葱段破開兩邊，用 6 湯匙油小火煎炸至出味，葱段不要，葱油留用。

4. 把留起的大葱段撕走三層外衣，只取葱黃（即葱芯），用 3 湯匙葱油慢火煎炸至微黃，取出。

5. 燒熱鑊中餘油，爆香薑片和蒜片，轉中火，放入蝦籽煸炒至香，潷酒。

6. 加入烏參、鹽、糖、老抽和蒸烏參的雞湯，煮沸，放入葱黃，煮至濃稠，用生粉勾芡，最後淋上 1 湯匙葱油，即可上碟。

雜菌海參鍋巴

Sea Cucumber and Mushrooms
over Crispy Rice Cakes

份量：4 人份 │ 準備時間：10 分鐘 │ 烹調時間：10 分鐘

建議採用

斯里蘭卡白玉沙禿參　馬達加斯加禿參　南美禿參　澳洲禿參

材料

麻油：1 茶匙	生粉：1 湯匙	糖：1/2 茶匙	鹽：1/2 茶匙	蠔油：1 湯匙	淡雞湯：250 毫升	蒜蓉：1 湯匙	鍋巴：6 塊	雜菌：300 克	發好海參：300 克

做法

1. 海參汆水，切成片或條狀。
2. 雜菌洗淨，汆水，瀝乾。
3. 鍋巴用中火炸脆，排在深碟中。
4. 燒熱 2 湯匙油，爆香蒜蓉，放入海參、雜菌和雞湯沸煮 2 分鐘，加蠔油、鹽和糖，用生粉勾芡，拌入麻油，上桌時趁熱倒在鍋巴上即成。

海參

英文食譜見第 163 頁

豆酥海參

Sea Cucumber with Soy Bean Crisp

份量：4人份 │ 準備時間：10分鐘 │ 烹調時間：5分鐘（不連煮豆酥時間）

建議採用

馬達加斯加禿參

斯里蘭卡白玉沙禿參

南美禿參

澳洲禿參

材料

發好海參：300克

豆酥：250毫升

蒜頭：2瓣，剁碎

乾葱：2粒，剁碎

幼鹽：1/2茶匙

胡椒粉：1/4茶匙

做法

1. 海參氽水，切片，排好在碟中。
2. 在鑊中放2湯匙油，用中火先把乾葱和蒜蓉炒香，放進豆酥，加鹽和胡椒粉，略為爆炒後鋪在海參上即成。

海參

英文食譜見第 164 頁

豆酥材料

黃豆：50克	水：375毫升	鹽：1/4茶匙	糖：1/4茶匙	頭抽：1/4茶匙	油：2湯匙

豆酥做法

1. 黃豆洗淨後用水泡4小時，再用攪拌機把黃豆連水攪碎成糜狀。

2. 用布把糜狀的黃豆水擠壓過濾，過濾的汁煮過即成豆漿。

3. 剩下的豆渣在易潔鑊上用慢火烘乾，也可用微波爐或烤爐烘乾。

4. 在豆渣裏加入鹽、糖和頭抽拌勻，徐徐加入油，用慢火煎炸至酥即成豆酥。

5. 豆酥可以一次過多做，用密封的玻璃瓶裝起，放在冰箱裏可長時間保存。

糟溜海參

Sea Cucumber in
Distilled Grain Sauce

份量：4人份 ｜ 準備時間：20分鐘 ｜ 烹調時間：10分鐘

建議採用

馬達加斯加
禿參

斯里蘭卡白玉
沙禿參

材料

發好海參：300克
無鹽雞湯：125毫升
乾雲耳：2克
筍片：50克
薑片：10克
糟滷：4湯匙
紹興酒：1湯匙
糖：2茶匙
生粉：2湯匙

做法

1. 發好的海參斜切厚片，汆水，用雞湯小火煮約10分鐘至入味。

2. 乾雲耳浸發後撕成小塊，用沸水灼熟。筍片汆水。

3. 燒熱1湯匙油，爆香薑片，放入筍片、雲耳、糟滷、紹酒、糖和250毫升水一起煮沸，然後放入海參連雞湯煮約5分鐘，用生粉水勾成濃芡，淋上1湯匙油，輕輕一拌即起。

海參

英文食譜見第165頁

乾燒海參

Sea Cucumber in Spicy Sauce

份量：4人份 │ 準備時間：10分鐘 │ 烹調時間：20分鐘

建議採用

斯里蘭卡白玉沙禿參

馬達加斯加禿參

南美禿參

澳洲禿參

南美刺參

材料

發好海參：約300克

小紅椒：2隻

榨菜：10克

郫縣豆瓣醬：1/2湯匙

乾雲耳：2克

冬菇：1朵

生粉：1湯匙

蒜蓉：1湯匙

薑蓉：1湯匙

半肥瘦絞豬肉：50克

紹興酒：1湯匙

頭抽：1茶匙

糖：2茶匙

陳醋：1湯匙

胡椒粉：1/2茶匙

麻油：1茶匙

葱：4條，切葱花

海參

英文食譜見第 166 頁

做法

1. 海參汆水 1 分鐘，用廚紙吸乾海參內外的水分。

2. 小紅椒去籽切粒，榨菜剁碎，郫縣豆瓣醬剁碎，雲耳浸透切碎，冬菇浸透去蒂切小粒。

3. 在炒鍋下油燒熱，先把海參抹上一層生粉再炸脆，撈出，瀝油。

4. 在鍋中燒熱 2 湯匙油，爆香蒜蓉和薑蓉，放入小紅椒和豆瓣醬炒香，加絞豬肉煸炒至酥香，灒酒，加榨菜、冬菇、雲耳、頭抽和糖，一起炒勻。

5. 加入 350 毫升水，煮沸。

6. 將海參放入鍋中，改中火燒煮約 2 分鐘，把海參轉身，再煮 2 分鐘，小心盛出海參放在碟中。

7. 把鍋中的湯汁用大火煮至稠濃，加入陳醋、胡椒粉和麻油煮沸，用生粉勾薄芡，淋在海參上，撒上葱花，即可上桌。

海參豆湯飯

Yellow Split Pea Soup with Rice and Sea Cucumber

份量：2-4 人份 ｜ 浸泡時間：8 小時 ｜ 烹調時間：45 分鐘

建議採用

馬達加斯加禿參

澳洲禿參

斯里蘭卡白玉沙禿參

南美禿參

材料

乾豌豆（馬豆）：100 克

發好海參：150 克

無鹽雞湯：350 毫升

薑汁：1 茶匙

生粉：2 湯匙

白米飯：1 碗

鹽：1/4 茶匙

做法

1. 豌豆用水沖淨，放在大碗裏加水泡 8 小時，瀝乾，用 250 毫升水煮 20 分鐘，再放入攪拌機攪拌成糊狀，用不黏鍋把豌豆糊炒至沒有多餘的水分。

2. 把發好的海參汆水，切粒，用 100 毫升雞湯加薑汁煨 10 分鐘，瀝乾。拌入生粉，炸脆。

3. 把豌豆糊加 250 毫升雞湯煮沸，加入適量的鹽調味，放入米飯煮 2-3 分鐘，裝入碗中，再放上海參即成。

海參

英文食譜見第 167 頁

蔘參相印

Double Boil Sea Cucumber and Ginseng

份量：2-4人份 ｜ 準備時間：5分鐘 ｜ 烹調時間：90分鐘

建議採用

黑石參

白石參　　　　　　　南美禿參

材料

發好豬婆參：300克

新鮮人蔘：1棵

無鹽雞湯：500毫升

去皮薑片：30克

去核紅棗：1粒

雞：半隻

鹽：適量

做法

1. 把海參和人蔘洗淨，分別汆水，瀝乾。

2. 把人蔘釀入海參內，放入燉盅，加入雞湯、雞、紅棗和薑，加蓋，燉90分鐘，加鹽調味即成。

海參

英文食譜見第167頁

竹笙海參排骨湯

Sea Cucumber, Bamboo
Fungus and Sparerib Soup

份量：4人份 │ 浸泡時間：8小時 │ 準備時間：5分鐘 │ 烹調時間：30分鐘

建議採用

北海道刺參　南美刺參　日本沖子刺參　大連刺參　日本關西刺參

材料

發好刺參：4條

北菇：4朵

冰鮮竹笙：4條

杞子：12粒

無鹽雞湯：500毫升

排骨：300克，斬件

薑汁：1湯匙

鹽：1/2茶匙

做法

1. 北菇浸8小時，去蒂，用生粉洗淨，再用清水洗去生粉。

2. 把海參和竹笙氽水1分鐘，瀝乾。

3. 排骨氽水1分鐘，瀝乾。

4. 杞子略為沖洗，瀝乾。

5. 煮沸雞湯，放入薑汁、排骨和北菇煮15分鐘，加海參、竹笙和杞子，同煮約15分鐘，再加鹽調味即成。

海參

英文食譜見第168頁

花膠和魚肚

花膠、魚肚和魚膠都是同品，並非魚的肚子，而是曬乾了的魚鰾。魚鰾是魚身上一個控制魚在水中升降的內臟器官，除了比目魚和鯊魚之外，大部分的魚類都有魚鰾。其中以黃花魚、白花魚、鱉魚、海鰻等大魚的魚鰾曬乾製成的，最富含膠質，故又俗稱為「膠」。潮州人叫藥用的魚肚做「魚膠」，作為烹調用的魚肚叫做「魚鰾」。

花膠一定是魚肚，但魚肚不一定是花膠。用黃花魚和白花魚的魚鰾曬乾製成的才叫做花膠，其他的都是魚肚，例如鱉肚（廣肚）、鴨泡肚、門鱔肚、赤魚膠、金山肚、鱔肚、蝴蝶肚，和上世紀五十年代之前，香港最名貴的大澳鱉肚，分得很清楚；不過，後來為了名字好聽，一般人都統稱之為花膠。

從中醫的角度看來，花膠甚具食療功效，它補而不燥，可主治身虛體弱，腎臟機能衰退，精稀薄少，又能長腰力，調理五臟機能，更有滋陰、固腎的功效，被女士視為養顏珍品。

從營養學角度分析，魚鰾（花膠或魚肚）的營養成分以蛋白質為主，當中含高黏性或生物性大分子膠原蛋白質、豐富骨膠原和黏多醣，佔總數高達86%，內含之磷質、鈣質、脂肪和鐵質，可助人體迅速消除疲勞，對外科手術病人傷口之恢復也有幫助。

無論是花膠或魚肚，乾貨的顏色金黃而呈半透明，硬實而沒有味道，很像一塊塑膠，一般分為海魚及淡水魚膠兩大類，價格及其身份，就以其罕有性、來源，和件頭大小厚薄來區分，品種及品質分為十多級，價格差距非常大。

家庭常用的花膠或魚肚，不需要買名貴的厚身大花膠，但也不要揀價錢太平宜的薄身貨；若療效低就是浪費柴火，煲完湯之後花膠就溶掉了。如果不是作為送禮用，我們建議自食煲湯，就選購來自南美地區的紮膠公，因肚長而窄，叫做「窄膠」，後俗稱「紮膠」。我們喜歡選購每司馬斤（600 克）有 12 至 14 隻（頭）的級數，每隻約 40 至 50 克，這種尺寸和厚度的紮膠，很適合煲湯及吃膠，食療功效不錯，而且價格並不太昂貴，適合喜歡經常吃花膠來補身美容的女士。

　　家庭發花膠或魚肚要注意，首先所有容器一定要洗得很乾淨，不能有油漬，花膠或魚肚遇油就容易溶解爛開。發好的花膠或魚肚，用清水沖洗乾淨，抹乾水分，切成兩三塊，放入三文治食物袋，擠出空氣，密封好，放入雪櫃的冰格中冷藏。建議一次過發好幾隻，隨時可以拿一兩隻出來烹調或燉湯，方便快捷。由於發好的花膠或魚肚，膠質很重，所以一定要沖洗和印乾水，而且每小袋最好只放 1 隻，以免黏在一起。

常見花膠魚肚品種

　　海味店常見的魚肚，有鰵肚（廣肚）、白花膠、毛鱨肚（午肚）、蝴蝶肚、紮膠、鱔肚、龍牙肚、魚雲肚、英國肚（非洲魚肚）、紐西蘭肚、鴨泡肚、蛇肚（陰陽肚）、葫蘆肚等等。

鴨泡肚

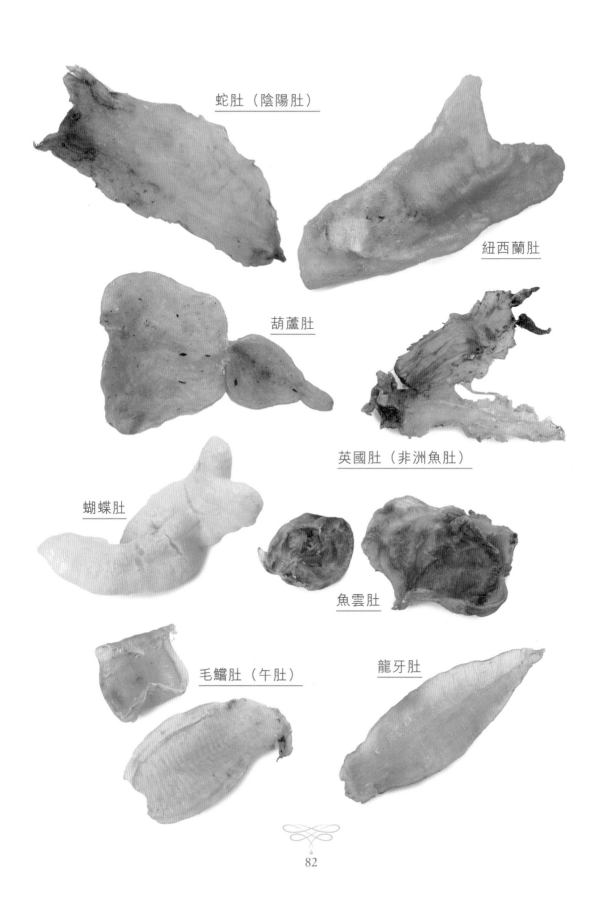

蛇肚（陰陽肚）

紐西蘭肚

葫蘆肚

英國肚（非洲魚肚）

蝴蝶肚

魚雲肚

毛鱔肚（午肚）

龍牙肚

花膠

　　花膠產自黃花魚和白花魚的魚鰾；因白花膠和黃花膠受季節性影響，供應量有限；所以形成品種珍貴，買少見少，其賣價亦是有上無下，而品質更是魚肚花膠中之極品。產地是中國沿海地區為主，但以黃海為多，南沙、西沙及印度洋也有出產，但數量不多。

浸發

1. 先煲薑水，水滾把乾的白花膠放下，熄火，浸焗 20 分鐘，撈起浸清水約 10-12 小時，白花膠便發脹一倍或以上；
2. 把白花膠切件，用清水盛着放入雪櫃浸 3-4 小時，即可煮食。

白花膠（產自中國黃海）

花膠和魚肚

✦ 鰵肚 （廣肚） ✦

　　鰵肚，別名廣肚或港肚，後者是地域性名稱。它源自鱈魚（俗稱鰵魚）的魚鰾，魚油豐富，故貯存多年後，會有磷光兼色澤變金黃。外表膠身大而寬闊，厚度十足，經烹調後仍保持堅挺而不瀉，主要產地來自印度洋，以巴基斯坦和印度沿岸最多。

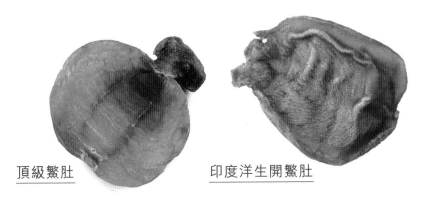

頂級鰵肚　　　　　　　印度洋生開鰵肚

浸發

1. 厚鰵肚最好放入雪櫃浸 2 天，期間每天換水 2 次（若不勤換水則會發出臭味，誤以為壞了）。

2. 煲大滾水，把已浸軟的鰵肚放下滾 20 分鐘，熄火焗 2 小時，撈起用清水浸 8-10 小時。此時，魚肚發大了 1 倍多。

3. 旁邊較薄的部位差不多已夠腍身，肚心因較厚而尚未發透還是實心，最好先把肚邊先切下來放入雪櫃用清水浸着。剩下的肚心再用大滾水煲 15 分鐘，又熄火焗 2 小時，再浸清水 8-10 小時。經過兩次浸、煲、焗，肚心便再浸發 1 倍大，便可以切件用清水浸着放入雪櫃。因魚肚是膠質食品，經過熱縮冷脹後，質感被冷水浸至結身，跌落地也彈得起，便算是夠身。這樣便完成浸發程式，可以隨意配用菜餚。

❖ 紮膠肚 ❖

紮膠肚是鱉肚的一種，原名「長肚」，因其肚形長而窄，所以又叫「窄膠」，「窄」字後來在海味行內寫成了「紮」字，一直沿用至今。其產地以中南美洲居多，如巴西、秘魯、委內瑞拉、古巴、巴拿馬、哥倫比亞、墨西哥等。

紮膠也分公乸，紮膠公的價錢比乸的貴。為甚麼花膠還要分公和乸？紮膠乸身形肥厚，外形特點是沒有直線紋，光滑透明，不耐火，容易溶於水，即俗語説的「瀉水」，口感滑「削」而軟糯。紮膠公的特別之處在於中間兩旁有修長的直線紋，俗稱「電車路」，外形修長，有兩隻「小耳朵」，光滑平面，頭闊尾窄，肉質不厚，比紮膠乸則多少許咬口（煙韌），也較為「見食」，用來煲湯不易溶於湯水，煲完湯還可吃有彈性口感的膠塊，據説藥效也較好。但是，紮膠公不宜在湯中煲太久，否則溶成糊狀，煲至腍身入味即可。所以先煲好靚湯，把浮油撇走，才後下發好的紮膠，在湯中煮 10 分鐘即可。

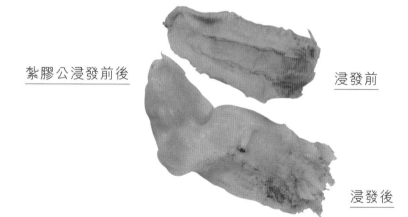

紮膠公浸發前後

浸發前

浸發後

花膠和魚肚

浸發（一）

先浸後焗的發鮃膠方法

1. 把鮃膠用清水浸泡1晚，浸至稍為軟身。
2. 在鍋裏放一大鍋冷水，放進薑片，把水燒至大滾，放入花膠，加蓋煮滾即熄火，中途不要打開，焗至水涼。
3. 取出鮃膠，重複上述2.的步驟兩次，直至鮃膠發脹1倍而且軟身。
4. 取出鮃膠，用清水沖洗，瀝乾水分即可煮食，或獨立包裝再冷凍貯存。

浸發（二）

先焗後浸的發鮃膠方法

1. 燒沸大鍋水，放下薑和葱，大火煮滾，放入鮃膠，加入1杯（250毫升）的米酒，改小火（攝氏95度左右），微滾20分鐘，熄火、加蓋，焗2小時，中途不要打開。
2. 撈出鮃膠，用冷水清洗，放入食物盒中，注入清水，放入冰箱中，浸泡兩至三天，每天必需換清水。完成後瀝乾水分，即可食用或獨立包裝再冷凍貯存。

浸發（三）

用壓力鍋發鮃膠

1. 把鮃膠用清水浸泡8小時或1晚，浸至稍為軟身。
2. 在壓力鍋中放進鮃膠和薑片，倒入清水，水要浸過鮃膠面約4厘米，蓋上鍋蓋。
3. 用低壓煲12分鐘，待放壓後即完成。如果是比較厚的鮃膠，可加多幾分鐘，至鮃膠夠軟身。放完壓應盡快撈出鮃膠浸入冰水中至涼，即可食用或獨立包裝再冷凍貯存。

鱔肚

鱔肚是海鰻魚、白門鱔魚的魚鰾，來自中國、孟加拉、印度洋、南美洲等地。發製鱔肚有鹽發、砂發和油發三種方法，所謂的砂爆魚肚即是用鹽發或砂發，採用炒熱的鹽或砂來焗發魚肚，魚肚表面色澤較暗啞；而油發即油炸，表面比較油亮，但比較油膩。但經用水煮及水焗處理後，三種魚肚的口感都差不多。

砂爆魚肚價格很平宜，烹調方便，不失為做家常菜的好材料。

浸發

1. 先把砂爆鱔肚浸軟，要以浮上水面為準。
2. 浸約 30 分鐘後，撈起切開為 1 吋長，用薑酒煲水滾 5 分鐘，便撈起過清水。用手擠出酒水把腥味帶走。擠乾後便可以用來製作各種菜式了。

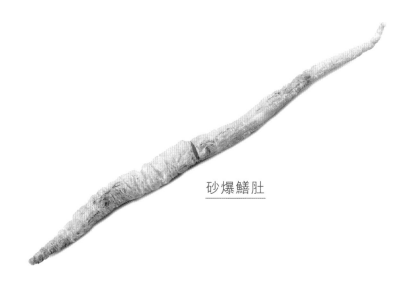

砂爆鱔肚

冬菇燴花膠

Braised Fish Maw
with Mushrooms

份量：4人份　│　準備時間：10分鐘　│　烹調時間：20分鐘

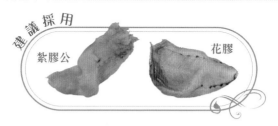

建議採用

 紫膠公　　　　花膠

材料

花膠1隻（未發重量）：60克

冬菇：8朵

蠔油：1湯匙

生粉：1/2湯匙

薑片：30克

清雞湯：250毫升

紹興酒：1湯匙

糖：1/2茶匙

鹽：1/2茶匙

生粉（勾芡用）：1/2茶匙

麻油：1/2茶匙

做法

1. 把花膠發好，取出，切成八塊。

2. 用水泡軟冬菇，去蒂，擠乾水分，拌入蠔油和生粉，再加入1茶匙油拌勻。

3. 燒熱1湯匙油，爆香薑片，放入冬菇、雞湯、酒、糖和鹽，煮沸，加入花膠，轉小火炆15分鐘，用生粉勾薄芡，拌入麻油即成。

花膠和魚肚

英文食譜見第169頁

魚肚棉花雞

Steamed Chicken
with Fish Maw

份量：4人份　準備時間：30分鐘　烹調時間：10分鐘

 材料

生粉：1湯匙

蒜頭：3瓣，剁蓉

糖：1/2茶匙

紹興酒：1/2湯匙

頭抽：1湯匙

蠔油：1茶匙

冬菇：4朵

薑汁：1湯匙

雞腿肉：400克

沙爆魚肚：40克

做法

1. 沙爆魚肚洗乾淨，放在1公升水內，加入薑汁，煮沸後加蓋熄火，浸焗30分鐘後取出，沖冷水，擠乾水分，切成約5厘米段。

2. 冬菇用清水泡軟，切除冬菇蒂，再切兩半，擠乾水，用蠔油拌勻，再拌入1/2湯匙油。

3. 雞腿切塊，拌入頭抽、酒、糖和蒜蓉，醃10分鐘，加入生粉拌勻。

4. 拌入魚肚、冬菇和1.5湯匙油，用大火蒸10分鐘即成。

花膠和魚肚

英文食譜見第170頁

瑤柱絲瓜燴魚肚

Braised Fish Maw with
Ridged Luffa

份量：4 人份 | 準備時間：35 分鐘 | 燉湯時間：20 分鐘

材料

沙爆魚肚：30克	江瑤柱：30克	薑汁：1湯匙	蝦米：10克	冬菇：4朵	紅蘿蔔：20克	絲瓜：300克	蒜頭：3瓣	頭抽：1湯匙	糖：1/2茶匙	清雞湯：125毫升	生粉：1/2茶匙

做法

1. 沙爆魚肚洗乾淨。用 1 公升水加薑汁煮沸，放入魚肚，煮沸後加蓋熄火，浸焗 30 分鐘後取出，沖冷水，擠乾水分，切成約 5 厘米段。
2. 江瑤柱撕成瑤柱絲（見第 105 頁），備用。
3. 蝦米用清水泡軟；冬菇用清水泡軟，切除冬菇蒂，再切兩半；紅蘿蔔切薄片；絲瓜刨去皮上的棱，滾刀斜切成塊。
4. 蒜頭去衣，用 1 湯匙油慢火煎香，加入魚肚、蝦米、冬菇、紅蘿蔔、頭抽、糖和雞湯，煮沸後繼續用慢火炆約 15 分鐘，再用生粉勾薄芡，最後撒上瑤柱絲即成。

花膠和魚肚

英文食譜見第 171 頁

瑤柱粟米魚肚羹

Fish Maw and Corn Soup

份量：4人份 ｜ 準備時間：30分鐘 ｜ 烹調時間：10分鐘

 材料

沙爆魚肚：100克

江瑤柱：30克

粟米蓉1盒／罐：418克

絞豬肉：150克

薑汁：1湯匙

清雞湯：250毫升

鹽：1茶匙

生粉：1/2茶匙

胡椒粉：1/2茶匙

做法

1. 魚肚洗乾淨，放進1公升水內，加入薑汁，煮沸後加蓋熄火，焗30分鐘。撈出魚肚，沖冷水，擠乾後切成小粒。
2. 江瑤柱撕成絲（見105頁），備用。
4. 絞豬肉碎加1/2茶匙鹽及生粉拌勻，備用。
5. 用鍋煮沸清雞湯，加入絞豬肉碎，用筷子把豬肉碎打散，放入瑤柱絲、粟米蓉及125毫升清水煮沸。
6. 加入魚肚粒同煮約3分鐘，加入1/2茶匙鹽和胡椒粉調味，即成。

花膠和魚肚

英文食譜見第172頁

花膠響螺燉雞湯

Double Boiled Chicken with Dried Fish Maw and Conch

份量：4人份 │ 準備時間：15分鐘 │ 燉湯時間：2小時15分鐘

建議採用

紫膠公

材料

花膠1隻：60克（未發乾品）

急凍響螺頭：250克

鹽：1茶匙

生粉：1茶匙

光雞：1/2隻（600克）

豬腱肉：150克

薑片：20克

紹興酒：1茶匙

熱開水：適量

鹽（調味用）：適量

做法

1. 浸發花膠，切件，備用。
2. 急凍響螺頭解凍後用鹽和生粉洗擦乾淨，切件，用沸水汆燙1分鐘，再過冷水。
3. 光雞用沸水汆燙1分鐘，過冷水。
4. 豬腱肉切成2厘米件，用沸水汆燙1分鐘，過冷水。
5. 把雞、豬腱肉、響螺件、薑片和酒放入燉盅中，加入適量熱開水至浸過材料，或至燉盅的大半容量，封蓋，大火燉2小時。
6. 把花膠用熱水稍為浸熱，放入燉盅，繼續燉15分鐘，取出，加鹽調味即成。

花膠和魚肚

英文食譜見第173頁

花膠腐竹糖水

Sweet Soup with Fish Maw and
Dried Bean Curd Sheets

份量：4人份 ｜ 準備時間：1小時 ｜ 烹調時間：30分鐘

材料

發好花膠：160克

腐皮碎：75克

生薏米：30克

白果：20粒

冰糖：60克

滾水：500毫升

做法

1. 薏米洗淨，加500毫升水和10克冰糖，煮沸，轉慢火煲60分鐘，瀝乾。

2. 白果洗淨，開邊，去芯，再清洗，加500毫升清水和10克冰糖，煮沸，慢火煮30分鐘，瀝乾。

3. 花膠切碎，汆水備用。

4. 把腐皮碎和40克冰糖放入滾水中，開中火，攪拌，煮約15分鐘，放入花膠，不斷攪拌，再煮約15分鐘至成牛奶狀，加入白果和薏米，煮沸即成。

花膠和魚肚

英文食譜見第 173 頁

江瑤柱

香港、澳門叫它做「江瑤柱」、「江珧柱」、「元貝」，台灣則一律稱為「乾貝」。

在廣東人的心目中，除鮑參翅肚外，江瑤柱在海味乾貨中，地位的重要性無可替代，是人們生活中最常食用的乾貨。

江珧柱和江瑤柱是來自兩種不同的貝類，一種是屬於江珧蛤科的江珧，另外一種是屬於海扇蛤科的扇貝。江珧柱是江珧的肌柱，主管殼的開合，牠的體型碩大而呈長形，頭尖尾寬像隻巨型帶子，柱肌超過整個體積的三分之一。江瑤柱則是扇貝的肌柱，扇貝是體型較小的圓蚌。兩種貝類的肌柱都可以做乾貝，把肌柱取出，經清洗、煮熟、烘乾後，放在陽光下晾乾，就成為乾貝，再按大小、色澤、形狀分為四級。野生的大江珧已越來越少，現在的瑤柱大部分都是自然放流養殖的扇貝。

大粒的日本江瑤柱，在粵菜中通常用來做瑤柱脯，是宴會中的傳統菜式。香港市場上出售的日本江瑤柱，主要是「宗谷貝」和「青森貝」兩種。北海道宗谷貝，是蝦夷扇貝肉柱做的乾貝，色澤金黃，品質最好。蝦夷扇貝即帆立貝，產於日本北海道及俄羅斯的冷水區。谷宗一帶的島嶼，坐落於日本北海道的極北海岸線上；一般來說，稚內、利尻島、禮文島和宗谷岬所出產或加工的江瑤柱，都被稱為「宗谷貝」。每年春天，來自北極的潔淨溶冰，為這片水域帶來了豐富的微生物，把帆立貝養得碩大肥美，所出產的「宗谷貝」是瑤柱中最優良等級。由於水質好，乾貝粒頭大而完整，味道濃郁，顏色較深而有光澤，價錢也最貴；在海味行中，宗谷不單只是一個地名，更是代表了優質的江瑤柱。2011 年 3 月 11 日，日本東北部海域發生九級地震，並引發了宮城和福島海岸的海嘯，海產業包括

「青森貝」，受到嚴重破壞和污染，自此「宗谷貝」獨領風騷，價格也成倍上升，但仍然深受高消費人群的歡迎。

中國沿海近年大量發展自然放流養殖的扇貝，生產江瑤柱和元貝成為具規模的海產加工業。青島出產的「青島赤貝」，元貝呈稍長形，顏色帶橙色，是市場的中價貨色，味道清甜，無渣，適合家庭用來煲湯、炆煮，以及炸瑤柱絲。

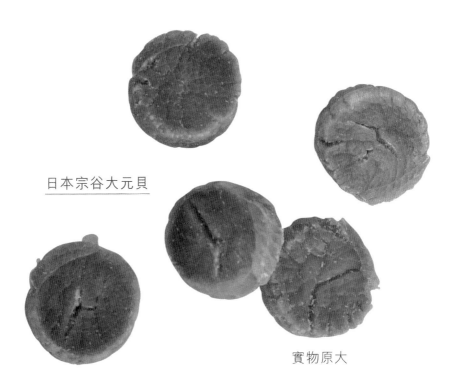

日本宗谷大元貝

實物原大

另一種越南出產的元貝，個子較小，鮮味不太濃，質感粗糙，鹹味重，浸洗後適合用來煲湯和煲粥。還有再次一等的元貝，叫做珠貝，是更細小的乾貝，鮮味淡，鹽味更重，但勝在價格平宜，浸洗後用來煲湯，作為提鮮也不錯。

江瑤柱含豐富鈣、碘、磷等人體所需的元素，用作菜餚可增加鮮味，煲湯或煲粥加入江瑤柱，更是滋補有益。江瑤柱和元貝是乾貨海味，比較耐放，分裝封好放在雪櫃冰格中更能保存鮮味；家中常備，就可以變化出精美的菜式。

浸發江瑤柱的方法

大粒宗谷江瑤柱，主要是做完整的瑤柱脯。把瑤柱排放在小碟中，用清水僅浸過面，浸 30 分鐘，再連水一起蒸 30 分鐘取出，小心把瑤柱上的韌帶撕去，即完成工序，原蒸汁可作烹調用。

細粒江瑤柱，一般是用來撕成瑤柱絲。用清水蓋過面浸 2 小時，去掉肉上韌帶，即完成浸發工序，如果是煲湯用，可連水放入湯鍋中一起煲。

越南元貝和珠貝，用清水沖淨後浸 30 分鐘，再用水洗一次即可用來煲湯，但浸的水鹽分很高，不宜留用。

炸瑤柱絲的方法

瑤柱浸發 30 分鐘後，去掉肉上韌帶，撕成絲，連水蒸 30 分鐘取出，用廚紙吸乾水分，攤開再晾乾，越乾越好。大火燒熱炸油至攝氏 215 度，一手拿鑊蓋，一手放下瑤柱絲，立即蓋上鑊蓋，熄火，炸 30 秒，開蓋立即用漏箕撈出，倒在準備好的吸油紙上，待紙上有油跡，再換紙，要重複換紙三四次，這是因為炸瑤柱絲很容易會再吸收油分而變軟。

瑤柱絲要炸得直而脆，就要用高溫油並熄火炸；但要注意的是，瑤柱絲下鑊遇油，會立刻四處飛彈起，必須立刻加鑊蓋，並穿着長袖衣服或護手套，小心被彈出的瑤柱絲燙傷，除非是專業食肆用，建議每次炸的份量不能多。

未炸的瑤柱絲　　　　　已炸的瑤柱絲

江瑤柱

發財瑤柱脯

Steamed Dried Scallops
with Black Moss

份量：6 人份 │ 準備時間：30 分鐘 │ 烹調時間：1 小時

材料

江瑤柱：50 克
髮菜：30 克
生菜：200 克
蒜頭：12 瓣
薑汁：1 茶匙
紹興酒：1 茶匙
頭抽：1 湯匙
糖：1 茶匙
清雞湯：1 杯（250 毫升）
蠔油：1 湯匙
生粉：1/2 茶匙
麻油：2 茶匙

做法

1. 髮菜用溫水浸過面，加幾滴油在水中，浸 1/2 小時後，換清水沖洗，撈出擠乾水分。
2. 100 毫升雞湯內加入頭抽、糖和 1 湯匙油，攪拌至糖完全溶化，再拌入髮菜中。
3. 把蒜頭去衣，炸至金黃色，撈出瀝油。
4. 江瑤柱用水稍為沖過，排在一個 14 至 15 厘米直徑的碗中，注入清水僅僅浸過面，浸 5 分鐘後，倒掉水，撕去邊上的硬枕不要，撕的時候要小心，把硬塊順着瑤柱的紋往下輕輕推，不要往外拉，否則會把瑤柱弄散。

江瑤柱

英文食譜見第 174 頁

5. 把炸過的蒜頭圍着瑤柱在碗中排好,加入薑汁和紹酒,注入清雞湯至浸過瑤柱面。

6. 最後放入髮菜,稍為壓實,用大火蒸 60 分鐘,取出。

7. 把生菜撕開,洗淨,用水焯熟,備用。

8. 把整個內有髮菜瑤柱的碗反扣在深碟上,圍上焯熟了的生菜。

9. 把餘下的雞湯煮沸,加入蠔油,用生粉開水埋芡,加入麻油煮沸,淋在瑤柱、髮菜和生菜上,即可上桌。

瑤柱桂花翅

Shark's Fin with Dried
Scallops and Eggs

份量：4人份　｜　準備時間：1小時15分鐘　｜　烹調時間：15分鐘

建議採用

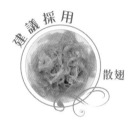

散翅

材料

江瑤柱：40克	發好散翅：120克	清雞湯：125毫升	銀芽：75克	雞蛋黃：6個	鹽：1/2茶匙

做法

1. 江瑤柱浸泡約 1 小時，瀝乾，撕成細絲，用廚紙吸乾水分，一半用油炸脆（見本書第 105 頁），另外一半留用。

2. 散翅用雞湯加 1/4 茶匙鹽煨至入味，瀝乾。

3. 銀芽汆水 5 秒鐘，過冷河，瀝乾水分。

4. 雞蛋黃加 1/4 茶匙鹽打勻。在鑊中把 1 湯匙油燒至中低溫，轉小火，放入蛋黃，用大勺底部沿着鑊底以打圈動作的把蛋黃壓成桂花狀。放入散翅和留用的瑤柱絲炒勻，再放入銀芽兜勻，盛出。

5. 把炸好的瑤柱絲撒在桂花翅上即成。

江瑤柱

英文食譜見第 175 頁

節瓜釀瑤柱脯

Stuffed Hairy Gourd with Dried Scallops

份量：6人份　準備時間：1小時　烹調時間：20分鐘

材料

宗谷瑤柱：6粒

節瓜：2條

清雞湯：125毫升

蠔油：1湯匙

生粉：1茶匙

麻油：1茶匙

江瑤柱

英文食譜見第 176 頁

做法

1. 瑤柱用水浸過面泡 30 分鐘，再蒸 30 分鐘，取出，瑤柱水留用。

2. 節瓜刮皮，切成 6 段，每一段要比瑤柱高 1/2 厘米，在每一段
   節瓜中挖出瑤柱大小的洞，但不要挖穿底部。

3. 把瑤柱釀在節瓜中，盛碟，大火蒸 20 分鐘，倒出碟中的水。

4. 把雞湯、瑤柱水加蠔油煮沸，用生粉勾薄芡，拌入麻油，淋在節
   瓜瑤柱上，即成。

江瑤柱

煎鹹魚瑤柱蒸肉餅

Steamed Minced Pork Pie with Salted Fish

份量：4人份 ｜ 準備時間：90分鐘 ｜ 烹調時間：10分鐘

材料

酶香馬友鹹魚：75克

江瑤柱：30克

絞瘦豬肉：250克

肥豬肉：100克

生粉：1湯匙

麥片：2湯匙

頭抽：1/2茶匙

鹽：1/2茶匙

糖：1茶匙

油：1湯匙

麻油：1/2茶匙

薑絲：5克

江瑤柱

英文食譜見第177頁

做法

1. 先把肥豬肉洗淨後用保鮮紙包着，平放在冰箱的冰格中，60分鐘後取出。這時的肥豬肉已稍為變硬，直刀切成小粒，再略為剁細。

2. 把絞碎的瘦豬肉再加剁一下，使肉粒更細。

3. 江瑤柱用5湯匙水浸軟，撕開成絲，浸江瑤柱的水留用。

4. 把絞瘦豬肉放在大碗中，加入頭抽、鹽、糖和浸過江瑤柱的水，拌匀後靜待15分鐘，讓調味品和水滲透到豬肉中。

5. 把絞豬肉用筷子順一個方向攪拌到起膠，然後用手把肉糰拿起用力撻回碗中，反覆撻十多下，使肉的表面起一層薄薄的膠質。

6. 把麥片用手捏成粉碎，與切好的肥豬肉粒一起拌匀，再拌入肉糰內。

7. 把瑤柱絲和生粉一同加到豬肉中拌匀，再加入油和麻油。

8. 把豬肉餅放在碟中，用手蘸水把肉餅的表面抹平，再包上微波爐保鮮紙，把肉餅密封，大火蒸約10分鐘。

9. 酶香鹹魚開慢火用少許油煎脆，放在蒸好的肉餅上，加上薑絲，即成。

雞蓉瑤柱羹

Dried Scallops and Chicken Soup

份量：4人份	浸泡時間：2小時	準備時間：15分鐘	烹調時間：10分鐘

材料

江瑤柱：30克
雞胸肉：150克
雞蛋白：1個
清雞湯：500毫升
鹽：1/2茶匙
胡椒粉：1/2茶匙
馬蹄粉：2湯匙
麻油：1茶匙

做法

1. 把江瑤柱浸泡2小時，撕成瑤柱絲。
2. 挑去雞胸肉的皮筋和膜，然後將雞胸肉剁爛成泥，放在大碗中。
3. 把蛋白拌入雞蓉中，用筷子打勻，挑出剩餘的筋膜不要。
4. 煮沸250毫升雞湯，趁熱倒入大碗中，同時不斷向同一方向攪動，至雞蓉半熟而完全溶入雞湯中。
5. 把其餘的雞湯煮沸，加入瑤柱和蒸瑤柱水煮3分鐘，加鹽和胡椒粉調味，用馬蹄粉開茨倒入攪至湯稠。
6. 改小火，把雞蓉湯徐徐倒下，用勺攪勻即熄火，加入麻油，盛入湯碗內。

江瑤柱

英文食譜見第178頁

蝦米、蝦乾和瀨尿蝦乾

蝦米

越南勾蝦米

大劍蝦米

馬來西亞金勾蝦米

泰國黃枝蝦米

　　蝦米，在菜式上也被稱為「金勾」、「開洋」。蝦米的做法，是把鮮蝦先煮熟，去殼，曬乾或烘乾而成，所以蝦米身呈捲曲狀。蝦米含豐富的蛋白質，味道鮮美，容易配搭菜式，是常用的家庭小乾貨。蝦米雖然比較耐放，但宜放在雪櫃中儲存。

　　市場上的蝦米，常見的有三種，主要產地來自越南、泰國、馬來西亞，以及中國東部沿海。市場上常見的蝦米品種很多，大劍蝦米來自浙江舟山，黃枝蝦米及麗蝦來自泰國，還有越南的勾蝦米。各地出產的蝦米，其鮮味和價錢相若，一般按蝦米的大小和顏色，以及菜式的需要來選購。

汕尾大紅蝦乾

✦ 蝦乾 ✦

蝦乾是把新鮮蝦去殼，在太陽下生曬而成，不經蒸煮，保持蝦身較大而直。蝦乾含豐富的蛋白質，味道鮮美，用來蒸、炒或做煲仔飯都很美味。

市場上的蝦乾，大部分產自中國東部沿海的福建漳州和廣東的潮汕地區，以及少數產自香港大澳的生曬九蝦乾，味道鮮甜，是香港的名特產；另外產自泰國的老虎蝦乾，體型較大，顏色鮮艷，價格也較貴。

體型比較細小的蝦乾用水浸約 15 分鐘，大隻蝦乾例如老虎蝦乾，需要浸 30 至 45 分鐘，浸至稍為軟身才可用，但勿浸太久因鮮味會減少。蝦乾容易受潮發霉，必需密封放在雪櫃中儲存。

泰國老虎蝦乾

✦ 瀨尿蝦乾 ✦

瀨尿蝦乾是把新鮮瀨尿蝦去殼，在太陽下生曬而成，過程不經蒸乾煮，蝦身軟薄而直。海味市場上的瀨尿蝦乾，大部分產自中國廣東的潮汕地區，也有來自泰國和越南，極少部分產自香港的大澳、長洲和屯門青山灣。香港出產的瀨尿蝦乾，味道最為鮮美，半乾濕的蝦身較柔軟，價錢也比較貴。

瀨尿蝦乾

瀨尿蝦乾含豐富的蛋白質，味道很鮮美，有的還會帶有蝦膏。瀨尿蝦乾浸 10 分鐘即可用，可用薑絲豉油蒸、做煲仔飯或配合炒蔬菜。瀨尿蝦乾容易受潮發霉，必需密封放在雪櫃中儲存。

瀨尿蝦乾蝦醬蒸肉片

Steamed Mantis Shrimps
and Pork with Shrimp Paste

份量：4人份　準備時間：10分鐘　烹調時間：10分鐘

材料

瀨尿蝦乾：25克

梅頭豬肉：150克

蝦醬：1/2茶匙

糖：1茶匙

薑汁：1/2湯匙

紹興酒：1/2湯匙

生粉：1茶匙

薑絲：5克

做法

1. 瀨尿蝦乾泡10分鐘，瀝乾。
2. 豬肉切片，用蝦醬、糖、薑汁、酒和生粉拌勻，盛碟，放上瀨尿蝦乾和薑絲，大火蒸10分鐘，淋上1湯匙熟油即成。

瀨尿蝦乾

英文食譜見第179頁

蝦乾魷魚肉片煲仔飯

Rice with Dried Shrimps, Dried Squids
and Pork in a Casserole

份量：4人份 | 準備時間：30分鐘 | 烹調時間：30分鐘

 材料

白米：320克

蝦乾：45克

半乾濕魷魚仔：30克

胸頭豬肉：150克

鹽：1/4茶匙

糖：1/4茶匙

頭抽：1/2茶匙

生粉：1/2茶匙

蒜蓉：1湯匙

做法

1. 白米洗淨，瀝乾。

2. 蝦乾泡水 30 分鐘，瀝乾。

3. 魷魚仔泡水 15 分鐘，瀝乾。

4. 豬肉切片，用鹽、糖、頭抽和生粉加 2 湯匙水拌勻，醃 15 分鐘。

5. 在鍋中燒熱 2 湯匙油，爆香蒜蓉，放入白米同炒 1 分鐘，加水覆蓋白米約 2 厘米，煮沸，開蓋，放入蝦乾、魷魚仔和肉片，大火煮至飯面收水，加蓋，轉小火焗約 15 分鐘即成

蝦乾

英文食譜見第 179 頁

土魷、墨魚乾和鱆魚

❧ 土魷 ❧

　　天然生曬的魷魚乾，香港稱為土魷，中國沿海的福建、廣東和台灣都有出產。在香港，大澳出產的土魷就最著名，用的魷魚來自香港及附近水域。土魷這個名字，據說這是因為海味商很講究意頭，水為財也，「乾」字表示無水無財，所以魷魚「干」的字體反轉，就成了「土」了，當然主要是本地出產的意思。

　　生曬土魷乾，分為全乾及半乾濕兩種。用原隻魷魚曬的土魷，叫做吊筒；另一種是剖開魷魚身來吊曬的，叫做吊片。香港的東部西貢一帶海域，出產一種魷魚叫做火箭魷，以前主要集中在九龍灣生曬成土魷乾，叫做「九龍吊片」，吊片薄身爽脆，味道鮮甜，是著名的香港特產。九龍灣本來是一個海灣，七十年代填海造地，發展成工業和住宅區，火箭魷吊片轉到大澳生曬，但仍沿用「九龍吊片」的名字。

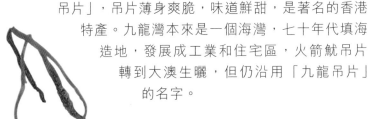

北海乾魷

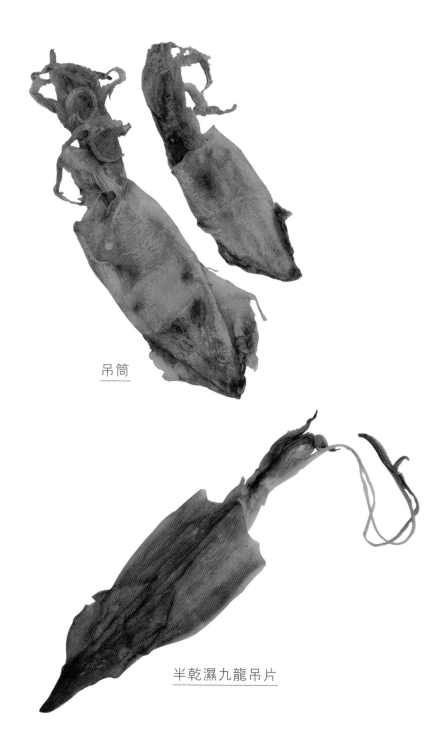

吊筒

半乾濕九龍吊片

土魷、墨魚乾和鱆魚

選購方法

　　全乾的土魷，要揀聞起來味香，手感肉厚，魷魚身上有一層白粉膜，乾爽不潮濕，就為上品。儲存土魷，要放在保持乾燥的地方，最好是密封放在雪櫃中，以免發霉。選購半乾濕的土魷，要揀手感柔軟，魷身乾淨無發霉，由於容易變壞，不宜儲存過多，而且必須密封存放在雪櫃。

浸發方法

　　全乾的土魷，魷身乾硬，必須經過浸發才可以食用。一般家庭浸發，可用 3 杯清水加 2 茶匙粗鹽拌勻，把土魷放入，浸泡 3 小時左右，取出用清水沖洗乾淨便可；這樣浸發的土魷，魷味較濃，但缺點是比較硬身，小孩老人吃時可能費勁。如果喜歡口感軟一些，可加少量食用小蘇打代替粗鹽，浸約 1 小時，之後再換清水浸 2 小時，浸清水中途要換兩次水，土魷會發大了，容易咀嚼，但味道就會較淡。

　　浸的時間長短，是根據土魷的大小和厚度來決定，也根據烹調所需來決定；要注意的是，浸的時間越長，土魷越鬆軟厚身，但鮮味越淡。如果要做到車仔麵那種水盆魷魚，酒家或食物工場的做法是用鹼水開水來浸，啤水至發脹，再用冷水浸 6 小時以上，所以味道很淡。

　　半乾濕土魷，魷身比較柔軟，但因為是曬前加了鹽來防腐，味道很鹹；做法是洗淨後浸水 20 至 30 分鐘，中途換水 1 次，主要是要減低鹹度。

浸發好的土魷，把外面的膜撕去，然後攤開魷魚，在肚內的那一面刈花。由於魷魚的纖維是橫着長的，在刈花後切成段，這樣煮熟時便會捲起成魷魚花；如果想要直身的土魷絲，例如做小炒的菜式，就要用刀順纖維橫切成幼條，這樣就捲不起來了。

✧ 墨魚乾 ✧

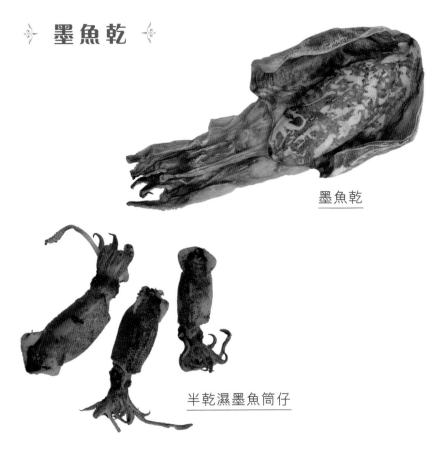

墨魚乾

半乾濕墨魚筒仔

墨魚又名烏賊，香港和中國沿海都有出產，墨魚乾是生曬墨魚，其中以浙江舟山群島出產最多，香港的離島有少量出產。墨魚乾的味道帶海水鹹味，香味濃郁，含豐富的蛋白質，多種維他命和鈣、磷、鐵等礦物質。食用方面，墨魚乾比不上土魷普及，一般用來煲湯或炆豬肉，味道香濃。

浸發方法

按墨魚乾的大小厚度，浸清水 15 分鐘，撕開去骨，再用清水浸約 10 分鐘，撈出瀝水即可，第二次浸的水可作煲湯或炆煮用。

❧ 鱆魚 ❧

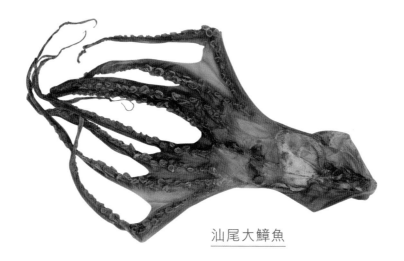

汕尾大鱆魚

　　鱆魚即八爪魚，而廣東人稱鱆魚的，就一定是指曬乾鱆魚，鮮活的鱆魚叫做八爪魚。香港市場上出售的曬乾鱆魚，全部產自中國沿海，包括山東、福建和廣東，最常見的是產自廣東汕尾的大鱆魚。

　　鱆魚味道鮮美，而曬乾的鱆魚更是香味濃郁，含豐富的蛋白質和礦物質。乾鱆魚煲湯除了鮮味香濃之外，更能滋補活血，是老幼咸宜的湯水，廣東人認為鱆魚煲湯還有利婦女產後補身，以及有催乳的功效。

浸發方法

　　用清水浸鱆魚 1 小時至軟身，即可用來煲湯。如果是用來蒸肉餅或做鱆魚雞粒炒飯，為免家中老人和小孩難以咀嚼，把鱆魚用清水浸軟後，剪成小塊，放在碟中，用水浸過鱆魚面，大火蒸 15 分鐘，取出再切成小粒，這樣鱆魚粒就會變腍了。

海參墨魚乾炆豬肉

Stewed Sea Cucumber with Dried
Cuttlefish and Pork

份量：4人份 │ 浸泡時間：2小時 │ 準備時間：10分鐘 │ 烹調時間：2小時

建議採用

馬達加斯加
禿參

澳洲禿參

斯里蘭卡白玉
沙禿參

南美禿參

❖ 材料 ❖

墨魚乾：150克

五花腩：600克

發好海參：250克

薑片：50克

麵豉醬（磨原豉）：1湯匙

紹酒：125毫升

頭抽：2湯匙

糖：1湯匙

鹽：1茶匙

生粉：1茶匙

墨魚乾

英文食譜見第 180 頁

做法

1. 墨魚乾泡 15 分鐘，撕去墨魚骨，再浸 10 分鐘，取出切成塊。

2. 五花腩汆水，切成 2 厘米方塊。

3. 海參汆水，切成 2 厘米件。

4. 鍋中燒熱 2 湯匙油，炒香薑片和麵豉醬，放入墨魚乾爆炒，放入五花腩，潸酒，加水覆蓋所有材料，煮沸，轉小火炆 1.5 小時。

5. 放入海參、頭抽、糖和鹽，炆 15 分鐘，大火收汁，用生粉勾薄芡即成。

鱆魚雞粒炒飯

Chicken Fried Rice
with Dried Octopus

份量：4人份 ｜ 準備時間：20分鐘 ｜ 烹調時間：5分鐘

材料

白飯：3碗	鱆魚：100克	雞腿肉：200克	薑米：1湯匙	葱花：2湯匙	頭抽：1湯匙

雞肉醃料

頭抽：1茶匙	糖：1/2茶匙	生粉：1茶匙

做法

1. 白飯用水沖洗，把部分澱粉洗去，用手弄散，瀝乾水分。

2. 用水浸過鱆魚面，大火蒸15分鐘，取出再切成小粒，用1茶匙生油拌勻。

3. 雞腿肉切粒，加醃料醃15分鐘。

4. 用2湯匙油起鑊，爆香薑米，放入雞腿肉、鱆魚粒同炒，再放進飯和頭抽同炒，加入葱花兜勻，即成。

鱆魚

英文食譜見第181頁

蓮藕鱆魚豬蹄湯

Dried Octopus Soup with
Lotus Roots and Pork

份量：4 人份 ｜ 準備時間：15 分鐘 ｜ 烹調時間：2 小時 45 分鐘

 材料

鱆魚：60 克	豬蹄：600 克	蓮藕：600 克	綠豆：50 克	薑片：15 克	清水：2.5 公升	鹽：1 茶匙

做法

1. 鱆魚洗淨，用溫水浸約 1 小時，備用。

2. 豬蹄洗淨，汆水瀝乾。

3. 蓮藕洗淨分節切開，每節切口切去約 1 厘米的節，清洗孔內，備用。

4. 綠豆用清水浸泡半小時，用筷子把綠豆塞進蓮藕的孔中，填至八成滿。

5. 用湯鍋煮沸約 2.5 公升水，放下薑片、豬蹄、蓮藕和鱆魚，大火加蓋沸煮 10 分鐘後，撇去浮沫，改中小火煮 2.5 小時，加鹽調味，即成。

鱆魚

英文食譜見第 182 頁

薑汁蒸香港仔魷筒

Steamed Dried Squids
with Ginger Juice

份量：4人份 │ 準備時間：30分鐘 │ 烹調時間：5分鐘

材料

香港仔魷筒：100克

薑汁：1湯匙

紹酒：1茶匙

糖：1茶匙

油：1湯匙

薑絲：10克

頭抽：1茶匙

葱花：1湯匙

做法

1. 香港仔魷筒洗淨，浸泡30分鐘，瀝乾。

2. 拌入薑汁、紹酒、糖和1湯匙油，放上薑絲，蒸5分鐘，淋上頭抽，撒上葱花即成。

土魷

英文食譜見第183頁

土魷蒸肉餅

Steamed Pork Pie with Dried Squid

份量：4人份 ｜ 浸泡加準備時間：2小時15分鐘 ｜ 烹調時間：10分鐘

 材料

全乾土魷：50克	馬蹄：2顆	肥豬肉：75克	絞脢頭瘦肉：200克	鹽：1/2茶匙	糖：1/2茶匙	頭抽：1/2湯匙	蒜蓉：1/2湯匙	生粉：1湯匙	胡椒粉：1/2茶匙	麥片：1湯匙	麻油：1茶匙

做法

1. 土魷洗淨後，平放，用少許暖水浸過面，浸2小時，撕去膜和骨，切成小粒。浸土魷水過濾，備用。
2. 馬蹄削皮，洗淨，切成小粒，泡在冷水中。
3. 肥豬肉洗淨，包起放在冰箱的冰格中凍至稍硬，切成小粒。
4. 把絞脢頭瘦肉加工再剁細一點。
5. 在脢頭瘦肉中加入2湯匙浸土魷水、鹽、糖和頭抽，拌勻，放10分鐘，再用筷子沿同一方向攪拌片刻至起膠。
6. 拌入土魷粒、肥豬肉粒、馬蹄粒和蒜蓉，再加入生粉及胡椒粉拌勻。
7. 加入揑碎了的麥片，再拌入麻油，放在蒸碟中，做成肉餅狀。
8. 把肉餅放入蒸鍋，大火隔水蒸10分鐘，即成。

土魷

英文食譜見第184頁

韭黃炒鴛鴦魷

Stir-fried Squid with Hotbed Chinese Chives

份量：4人份 ｜ 準備時間：40分鐘 ｜ 烹調時間：5分鐘

材料

鮮魷1至2隻：約500克

半乾濕九龍吊片：1隻

薑汁：1湯匙

韭黃：約150克，切5厘米段

蒜頭：4瓣，剁蓉

乾葱頭：兩粒，切成4瓣

鹽：1茶匙

糖：半茶匙

紹興酒：1茶匙

麻油：1/4茶匙

冰水：1大碗

生粉：1/2茶匙

土魷

英文食譜見第 185 頁

做法

1. 鮮魷切開膛，撕去外層薄膜和附翼不要，只用魷魚身，從中間順長度切開成兩邊。

2. 在砧板上鋪一塊廚紙，把一邊魷魚，腹內向上平放紙上，左手輕按魷魚用斜刀交叉剖上花紋（見第 131 頁）。剖完花後，再橫切成約 6 厘米長的段。以同樣方法處理另一邊魷魚。

3. 半乾濕九龍吊片洗淨後浸水 30 分鐘，中途換水 1 次，主要是減低鹹度，沖洗乾淨，按鮮魷的處理方法準備乾魷。

4. 用鑊燒一大鍋水，加薑汁，左手持一漏勺，右手倒入九龍吊片，立即熄火，右手用筷子散開吊片，余約 3 秒左右，立即用漏勺撈出，瀝去水分，倒在大碗冰水中，浸到水溫不再冷，撈出瀝水。

5. 把水再燒開，左手持一漏勺，右手倒入鮮魷魚塊，立即熄火，右手用筷子散開魷魚，余約 5 秒左右，立即用漏勺撈出魷魚，瀝去水分，倒在大碗冰水中，浸到水溫不再冷，撈出魷魚瀝水。

6. 大火燒開 2 湯匙油至七成熱，爆香乾蔥和蒜蓉，倒入魷魚和吊片加鹽爆炒數下，加入酒、糖和韭黃快速同炒，加入麻油，埋薄芡即成。

虎蝦乾土魷炒韭菜花

Stir Fried Dried Squid, Dried Prawns
with Flowering Chinese Chives

份量：4人份 ｜ 準備時間：30分鐘 ｜ 烹調時間：3分鐘

 材料

胡椒粉：1/8茶匙

頭抽：1茶匙

紹興酒：1湯匙

沙茶醬：1茶匙

紅甜椒：1隻

韭菜花：200克

五花腩：150克

虎蝦乾：30克

半乾濕吊片：1隻（約50克）

做法

1. 吊片泡水30分鐘，撕去外層的膜，平放在砧板上，橫切成粗絲，備用。

2. 虎蝦乾用水僅浸過面，泡20分鐘，連水蒸10分鐘，取出蝦乾。

3. 五花腩汆水，切成薄片。

4. 韭菜花洗淨切段，紅椒切粗條。

5. 燒熱1湯匙油，爆香吊片絲、虎蝦乾、五花腩和紅椒，然後加入沙茶醬兜勻及在鑊邊灒酒，再加入韭菜花，大火同炒至收乾汁，最後加頭抽及胡椒粉炒勻即成。

土魷

英文食譜見第 186 頁

Contents

Abalone and Sea Cucumber Salad with Fish Flavored Sauce

*Chinese version refer to p.32

INGREDIENTS
300-400 g sea cucumber, pre-soaked
2-3 canned abalones
1 tbsp chopped ginger
1 tbsp chopped garlic
1/2 tbsp chopped Pixian chili paste
2 sprigs spring onion, shredded

SAUCE
1 tbsp Shaoxing wine
1 tbsp sugar, 1 tbsp black vinegar
1/2 tsp top soy sauce
1/4 tsp salt
1/2 tsp white pepper
1 tsp sesame oil

METHOD
1. Blanch sea cucumber for 1 minute, drain, and cut into 5x1 cm strips.

2. Cut abalones into similar size as sea cucumbers.

3. Heat 2 tbsp of oil, stir fry ginger, garlic and chili paste, add sauce and stir fry into fish flavored sauce. Put sauce into a plate, add sea cucumbers and abalones, and top with shredded spring onions.

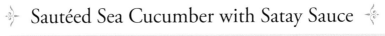 Shrimp and Sea Cucumber Salad with Chili Pepper Dressing

Serving: for 4 | Preparation time: 15 minutes | Cooking time: 15 minutes

*Chinese version refer to p.36

INGREDIENTS

300 g sea cucumber (pre-soaked), 300 g shrimps, 4 Hunan chili peppers
1 tbsp lemon juice, 1/4 tsp salt, 1/2 tbsp sugar,
1/2 tsp crushed black pepper, 1 tbsp olive oil

METHOD

1. Blanch sea cucumber, cool, and cut to 2 cm pieces.

2. Shell shrimps, de-vein, blanch, drain and set aside to cool.

3. De-seed chili peppers, and roast over fire until the skins are half burnt. Chop.

4. Mix chili pepper, lemon juice, black pepper, salt, sugar and olive oil into a chili pepper sauce to serve as salad dressing. Add sea cucumber and shrimps.

Sautéed Sea Cucumber with Satay Sauce

Serving: 4 | Preparation time: 15 minutes | Cooking time: 3 minutes

*Chinese version refer to p.40

INGREDIENTS

300 g sea cucumber (pre-soaked) , 150 g Chinese broccoli,
2 cloves garlic (sliced), 1 tbsp satay sauce, 1/2 tsp sugar,
1 tbsp Shaoxing wine, 1/2 tsp corn starch

METHOD

1. Blanch sea cucumber, drain, and cut into 5x1 cm lengths.

2. Rinse Chinese broccoli, cut into 5 cm sections and blanch.

3. Heat 2 tbsp of oil, brown garlic, add satay sauce, sugar and sea cucumber and sauté over high heat. Sprinkle wine and put in Chinese broccoli and stir fry, thicken with corn starch.

Prawns and Sea Cucumber with Gongbao Sauce

Serving: for 4   Preparation time: 20 minutes   Cooking time: 15 minutes

*Chinese version refer to p.42

INGREDIENTS

250 g sea cucumber, pre-soaked
8 medium prawns
2 tbsp corn starch
1-2 red chili pepper
1/2 each of red and green sweet pepper, cut into wedges
1 tsp Sichuan pepper
5 g ginger slices
10 g garlic slices
2 tbsp chopped spring onion
30 g shelled peanuts

FLAVORING SAUCE

1/4 tsp white pepper
1/2 tsp top soy sauce
1/2 tbsp vinegar
1/2 tsp corn starch
1/4 tsp salt
1 tsp sesame oil
1 tsp red oil

METHOD

1. Blanch sea cucumber, drain, and cut into 1.5 cm pieces.

2. Shell prawns, cut open at the back, de-vein, rinse and drain.

3. Deep fry peanuts, drain.

4. Mix sauce ingredients in a bowl, and stir well before use.

5. Mix sea cucumber with corn starch, and deep fry. Remove to drain oil.

6. Heat 1 tbsp of oil, stir fry red chili pepper, Sichuan pepper, sweet pepper, ginger, garlic and spring onion, put in prawns and sea cucumber and stir fry until prawns are fully cooked. Add flavoring sauce and sauté until sauce thickens. Mix in peanuts.

Sea Cucumber

Braised Sea Cucumbers with Beijing Scallions

*Chinese version refer to p.44

INGREDIENTS
4 sea cucumbers with spike, pre-soaked
125 ml unsalted chicken broth
1 tbsp ginger juice
2 Beijing scallions
2 tbsp oyster sauce
1 tsp sugar
1/8 tsp white pepper
1/2 tbsp corn starch
1 tsp sesame oil

METHOD

1. Rinse sea cucumbers, put into chicken broth and ginger juice and steam for 15 minutes.

2. Cut scallion stems into 5 cm sections, and save one half for later. Cut each of the remaining sections vertically into two halves. Heat 6 tbsp of oil and pan fry the scallions over low heat until dark brown. Discard scallions and save the scallion flavored oil.

3. Heat 3 tbsp of scallion flavored oil and pan fry the reserved scallion sections to golden brown. Remove scallions and add sea cucumbers together with the chicken broth, and bring to a boil.

4. Add oyster sauce, sugar and white pepper, and braise for 5 minutes, add scallions, and thicken with appropriate quantity of corn starch. Stir in sesame oil, transfer to plate, and drizzle 1 tbsp of scallion flavored oil on top.

Dried Scallop, Sea Cucumber and Fish Maw

Serving: for 1   Preparation time: 30 minutes   Cooking time: 45 minutes

*Chinese version refer to p.46

INGREDIENTS
1 dried scallop
1 spiky sea cucumber, pre-soaked
1 fish maw, pre-soaked
1 tsp ginger juice
1 tbsp Shaoxing wine
1 tsp oyster sauce
250 ml unsalted chicken broth
1/2 tbsp corn starch
1/2 tsp sesame oil

METHOD
1. Submerge dried scallop in water for 30 minutes and steam another 30 minutes.

2. Put scallop, fish maw and sea cucumber in a deep plate and add chicken broth, ginger juice and wine, and steam for 30 minutes. Pour out chicken broth.

3. Bring chicken broth to a boil in a wok, add oyster sauce, and thicken with corn starch. Stir in sesame oil and drizzle over scallop, fish maw and sea cucumber.

Braised Sea Cucumber with Pig Trotters

Serving: for 4 | Preparation time: 15 minutes | Cooking time: 2 hours

*Chinese version refer to p.50

INGREDIENTS
500 g pig trotters
300 g sea cucumber, pre-soaked
50 g ginger slices
4 stalks spring onion, sectioned
2 tbsp Shaoxing wine
250 ml unsalted chicken broth
1 tbsp top soy sauce
1 tbsp sugar
1/2 tsp salt
1 tbsp spring onion oil

METHOD
1. Blanch pig trotters, drain, and boil for 1.5 hours, rinse with cold water, de-bone, and cut trotters into 2.5 cm squares.

2. Blanch sea cucumber, drain, and cut into pieces similar in size to pig trotters.

3. Heat 1 tbsp of oil, and stir fry ginger and spring onions. Put in pig trotters and sea cucumber, drizzle wine, add chicken broth and bring to a boil. Reduce to low heat, add soy sauce, sugar and salt, and braise for 15 minutes.

4. Change to high heat to reduce sauce, and stir in spring onion oil.

Sweet and Sour Sea Cucumber

Serving: for 4   Preparation time: 5 minutes   Cooking time: 25 minutes

*Chinese version refer to p.52

INGREDIENTS
300 g sea cucumber, pre-soaked
125 ml unsalted chicken broth
3 tbsp corn starch
1 small can of pineapple
1/2 red sweet pepper
1/2 green sweet pepper
1/2 onion
2 cloves garlic, sliced
4 tbsp red vinegar
4 tbsp red sugar
1 tsp sesame oil

METHOD
1. Cut sea cucumber into chunks, braise with chicken broth for 10 minutes, drain, mix with corn starch and deep fry until crispy.

2. Deseed sweet peppers and cut peppers, onion and pineapple into chunks.

3. Blend vinegar and sugar into sweet and sour sauce.

4. Heat 2 tbsp of oil, stir fry garlic and onion, add sweet and sour sauce and cook until thicken. Put in sea cucumber and sweet peppers, and stir fry over high heat for about 1 minute. Thicken appropriately with corn starch and stir in sesame oil.

Sautéed Sea Cucumber with Pork

Serving: for 4 | Preparation time: 10 minutes | Cooking time: 3 minutes

*Chinese version refer to p.54

INGREDIENTS

300 g sea cucumber, pre-soaked
125 g pork
125 g flowering Chinese chives
1/4 tsp salt
1/4 tsp sugar
1/2 tsp corn starch
2 cloves garlic, sliced
1 tbsp oyster sauce
1/2 tsp corn starch for thickening
1 tsp sesame oil

METHOD

1. Cut pork into slices, and marinate with salt, sugar and 2 tbsp of water for 10 minutes. Put in corn starch and mix well.

2. Cut sea cucumber into slices and blanch for 1 minute. Drain.

3. Rinse flowering Chinese chives and cut into 5 cm sections.

4. Heat 2 tbsp of oil, stir fry garlic, add pork and sea cucumber, and stir fry for about 1 minute. Put in oyster sauce and 50 ml of water, boil, add flowering Chinese chives and sauté until done. Thicken with corn starch and stir in sesame oil.

Stuffed Sea Cucumbers

Serving: for 4   Preparation time: 30 minutes   Cooking time: 15 minutes

*Chinese version refer to p.56

INGREDIENTS

4 spiky sea cucumber, pre-soaked
150 ml unsalted chicken broth
1 tbsp ginger juice
300 g fresh shrimps
1/2 tsp coarse salt
10 g pork fatback, cut into bits
1/2 egg white
1/8 tsp salt, 1/8 tsp sugar
1/8 tsp white pepper
1/4 tsp corn starch

SPRING ONION SAUCE

6 stalks spring onions
1 tsp wasabi, 1/8 tsp salt

METHOD

1. Blanch sea cucumber, and braise with chicken broth and ginger juice for 30 minutes. Drain and reserve the chicken broth.

2. Shell shrimps and de-vein, put in a strainer, add coarse salt and rub shrimps thoroughly by hand. Rinse with cold water and drain. Clean and dry the chopping board, flatten shrimps on the chopping board with the body of a large kitchen chopper, and chop repeatedly with the blunt edge of the knife into a paste. Put paste into a large bowl.

3. Add egg white, stir vigorously with chopsticks in a single direction until gluey, mix in table salt, sugar, white pepper and corn starch, then pick up paste by hand and smash against the bowl a number of times until a gluey shrimp patty is formed.

4. Mix in pork fatback to make a filling, and stuff into the sea cucumber. Steam over high heat for 10 minutes.

5. Use only spring onion greens, blanch for 10 seconds, cool down in iced water, cut into sections, and blend with wasabi and chicken broth to make a sauce. Drizzle sauce over sea cucumbers.

Sea Cucumber

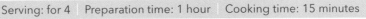# Braised Sea Cucumber with Shrimp Roe and Beijing Scallions

Serving: for 4 | Preparation time: 1 hour | Cooking time: 15 minutes

*Chinese version refer to p.60

INGREDIENTS

1 large sea cucumber, pre-soaked
250 ml unsalted chicken broth
2 tbsp shrimp roe
4 Beijing scallions
30 g ginger slices
2 cloves garlic, sliced
2 tbsp Shaoxing wine
1 tbsp sugar
1/2 tsp salt
1 tbsp dark soy sauce
corn starch as needed

METHOD

1. Put sea cucumber in chicken broth and steam for 30 minutes or until sufficiently soft.

2. Roast shrimp roe in a dry wok over low heat.

3. Cut Beijing scallion stems into 5 cm sections, and save one half for later. Cut each of the remaining sections vertically into two halves. Heat 6 tbsp of oil and pan fry the scallions over low heat until dark brown. Discard scallions and save the scallion flavored oil.

4. Remove 2 to 3 outer layers of the remaining scallions to expose the heart, and pan fry in 3 tbsp of scallion flavored oil over low heat until slightly brown. Remove from wok.

5. Heat the remaining oil in the wok, stir fry ginger and garlic until pungent, change to medium heat and stir in shrimp roe. Drizzle wine.

6. Add sea cucumber, salt, sugar, dark soy sauce and the chicken broth, bring to a boil, put in Beijing scallions, braise until sauce begin to thicken. Thicken sauce further with corn starch. Finally stir in 1 tbsp scallion flavored oil.

Sea Cucumber and Mushrooms over Crispy Rice Cakes

Serving: for 4   Preparation time: 10 minutes   Cooking time: 10 minutes

*Chinese version refer to p.62

INGREDIENTS

300 g sea cucumber, pre-soaked
300 g mixed mushrooms
6 rice cakes
1 tbsp chopped garlic
250 ml chicken broth
1 tbsp oyster sauce
1/2 tsp salt
1/2 tsp sugar
1 tbsp corn starch
1 tsp sesame oil

METHOD

1. Blanch sea cucumber and cut into slices or strips.

2. Clean and blanch mushrooms. Drain.

3. Deep fry rice cakes over medium heat until crispy and transfer to a deep plate.

4. Heat 2 tbsp of oil, stir fry garlic, add sea cucumber, mushrooms and chicken broth, and braise for 2 minutes. Add oyster sauce, salt and sugar, and thicken with corn starch. Stir in sesame oil and pour over crispy rice cakes.

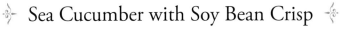

Sea Cucumber with Soy Bean Crisp

*Chinese version refer to p.64

INGREDIENTS
300 g sea cucumber, pre-soaked
250 g soy bean crisp
2 cloves garlic, chopped
2 shallots (chopped), 1/2 tsp table salt
1/4 tsp white pepper

METHOD
1. Blanch sea cucumber, cut into slices and arrange in a plate.
2. Heat 2 tbsp of oil in a wok, stir fry shallot and garlic until aromatic, stir in soy bean crisp, salt and white pepper, and put over sea cucumber.

Soy Bean Crisp

INGREDIENTS
50 g soy beans
375 ml water, 1/4 tsp salt
1/4 tsp sugar
1/4 tsp light soy sauce
2 tbsp oil

METHOD
1. Wash and soak soy beans in water for 4 hours, put soy beans together with the water into a blender and grind until they become a soy bean paste.
2. Filter the soy bean paste through a piece of cloth to separate the dregs from the milk. The milk can be boiled to make soy bean milk.
3. Roast the soy bean dregs in a non-stick pan over low heat to remove most of the moisture. It can also be done using a microwave oven or a conventional oven.
4. Mix salt, sugar and soy sauce with the soy bean dregs in the pan, gradually add oil and pan fry over low heat until crispy.
5. Soy bean crisps can be kept for a long time if refrigerated.

Sea Cucumber

Sea Cucumber in Distilled Grain Sauce

Serving: for 4   Preparation time: 20 minutes   Cooking time: 10 minutes

*Chinese version refer to p.68

INGREDIENTS

300 g sea cucumber, pre-soaked
125 ml unsalted chicken broth
2 g dried black fungus
50 g bamboo shoot slices
10 g ginger slices
4 tbsp distilled grain sauce
1 tbsp Shaoxing wine
2 tsp sugar
2 tbsp corn starch

METHOD

1. Slant cut sea cucumber into thick slices, blanch, drain, and simmer in unsalted chicken broth for about 10 minutes.

2. Soak black fungus until soft, tear into smaller pieces and blanch. Blanch bamboo shoot.

3. Heat 1 tbsp of oil and stir fry ginger until pungent. Add bamboo shoot, black fungus, distilled grain sauce, wine, sugar and 250 ml water, and bring to a boil. Add sea cucumber with chicken broth and cook for 5 minutes, mix corn starch with water and thicken sauce. Stir in 1 tbsp oil.

Sea Cucumber in Spicy Sauce

Serving: for 4 | Preparation time: 10 minutes | Cooking time: 20 minutes

*Chinese version refer to p.70

INGREDIENTS

300 g sea cucumber, pre-soaked
2 red chili peppers
10 g preserved mustard tuber
1/2 tbsp Pixian chili paste
2 g dried black fungus
1 dried black mushroom
1 tbsp corn starch
1 tbsp chopped garlic
1 tbsp chopped ginger
50 g minced pork
1 tbsp Shaoxing wine
1 tsp top soy sauce, 2 tsp sugar
1 tbsp black vinegar
1/2 tsp white pepper
1 tsp sesame oil
4 stalks spring onion, chopped

METHOD

1. Blanch sea cucumber for 1 minute, drain, and pat dry with kitchen towels.

2. Deseed and chop chili peppers, chop mustard tuber and chili paste. Soak dried black fungus until soft and tear into smaller pieces. Soak mushroom until soft, remove stem, and chop into small bits.

3. Heat oil in a wok, coat sea cucumber with corn starch and deep fry until crispy.

4. Heat 2 tbsp of oil, stir fry garlic and ginger, add chili peppers and chili paste, put in minced pork, sprinkle wine, and add mustard tuber, mushroom, black fungus, soy sauce and sugar. Stir fry to mix all ingredients.

5. Add 350 ml boiling water and re-boil.

6. Put in sea cucumber, reduce to medium heat and braise for 5 minutes. Turn over sea cucumber and braise for another 5 minutes. Transfer to plate.

7. Heat sauce over high heat until thicken, add vinegar, white pepper and sesame oil, re-boil, thicken further with corn starch and drizzle on the sea cucumber. Top with chopped spring onions.

Sea Cucumber

Yellow Split Pea Soup with Rice and Sea Cucumber

Serving: for 2-4   Preparation time: 8 hours   Cooking time: 45 minutes

*Chinese version refer to p.74

INGREDIENTS
100 g dried yellow split peas, 150 g sea cucumber (pre-soaked)
350 ml unsalted chicken broth, 1 tsp ginger juice, 2 tbsp corn starch
1 bowl steamed rice, 1/4 tsp salt

METHOD
1. Rinse peas and soak in a large bowl for 8 hours, drain, add 250 ml water and boil for 20 minutes. Blend in a food processor into a mush and remove excess moisture using a non-stick pan.
2. Blanch sea cucumber, dice, and simmer in 100 ml chicken broth and ginger juice for 10 minutes. Drain, mix in corn starch and deep fried until crispy.
3. Add split pea mush to 250 ml chicken broth, bring to a boil and flavor with salt. Put in steamed rice, cook for 2 to 3 minutes, and transfer to bowls. Top with sea cucumber.

 # Double Boil Sea Cucumber and Ginseng

Serving: for 2-4   Preparation time: 5 minutes   Cooking time: 90 minutes

*Chinese version refer to p.76

INGREDIENTS
300 g sea cucumber (pre-soaked), 1 fresh ginseng, 500 ml unsalted chicken broth
30 g ginger slices (peeled) , 1 jujube (stoned) , 1/2 chicken, salt

METHOD
1. Clean and blanch sea cucumber and ginseng separately, drain.
2. Stuff ginseng into the sea cucumber, put into a large bowl together with chicken broth, chicken, jujube and ginger, cover, and double boil for 90 minutes. Flavor with salt.

Sea Cucumber, Bamboo Fungus and Sparerib Soup

Serving: for 4 | Soaking time: 8 hours | Preparation time: 5 minutes | Cooking time: 30 minutes

*Chinese version refer to p.78

INGREDIENTS
4 spiky sea cucumbers, pre-soaked
4 dried black mushrooms
4 chilled bamboo fungus
12 wolfberries
500 ml unsalted chicken broth
300 g spareribs
1 tbsp ginger juice
1/2 tsp salt

METHOD
1. Soak mushrooms for 8 hours, cut off stems, and rub with corn starch to clean. Rinse.

2. Blanch sea cucumbers and bamboo fungus for 1 minute, drain.

3. Blanch spareribs, drain.

4. Rinse wolfberries, drain.

5. Bring chicken broth to a boil, put in ginger juice, spareribs and mushrooms, cook for 15 minutes, add sea cucumbers, bamboo fungus and wolfberries, boil for about 15 minutes, and flavor with salt.

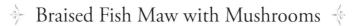# Braised Fish Maw with Mushrooms

Serving: for 4   Preparation time: 10 minutes   Cooking time: 20 minutes

*Chinese version refer to p.88

INGREDIENTS
60 g dried fish maw
8 dried back mushrooms
1 tbsp oyster sauce
1/2 tbsp corn starch
30 g ginger slices
250 ml clear chicken broth
1 tbsp Shaoxing wine
1/2 tsp sugar
1/2 tsp salt
1/2 tsp corn starch (for thickening)
1/2 tsp sesame oil

METHOD
1. Prepare fish maw and cut into 8 pieces.
2. Soften mushrooms in water, remove stems, and squeeze to remove the water. Mix mushrooms with oyster sauce and corn starch, and stir in 1 tsp of oil.
3. Heat 1 tbsp of oil, stir fry ginger until aromatic, add mushrooms, chicken broth, wine, sugar and salt, and bring to a boil. Put in fish maw, reduce to low heat and braise for 15 minutes. Thicken with corn starch and stir in sesame oil.

Steamed Chicken with Fish Maw

Serving: for 4   Preparation time: 30 minutes   Cooking time: 10 minutes

*Chinese version refer to p.90

INGREDIENTS

40 g dried fish maw
400 g boneless chicken thigh
1 tbsp ginger juice
4 dried black mushrooms
1 tsp oyster sauce
1 tbsp top soy sauce
1/2 tbsp Shaoxing wine
1/2 tsp sugar
3 cloves garlic, chopped
1 tbsp corn starch

METHOD

1. Rinse fish maw, put into 1 litre of water with ginger juice, re-boil, cover pot and turn off heat. Remove fish maw after steeping for 30 minutes, rinse with cold water, squeeze dried and cut into 5 cm sections.

2. Soften dried mushrooms with cold water and remove stems. Cut mushrooms in halves, squeeze dry, mix with oyster sauce, and then add 1/2 tbsp of oil.

3. Cut chicken into chunks, and marinate with soy sauce, wine, sugar and garlic for 10 minutes. Mix in corn starch.

4. Add fish maw, mushrooms and 1.5 tbsp of oil, mix well, and steam over high heat for 10 minutes.

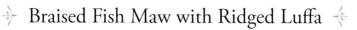

Braised Fish Maw with Ridged Luffa

*Chinese version refer to p.92

INGREDIENTS

30 g dried fish maw
30 g dried scallops
1 tbsp ginger juice
10 g dried shrimps
4 dried black mushrooms
20 g carrot
300 g ridged luffa
3 cloves garlic
1 tbsp top soy sauce
1/2 tsp sugar
125 ml clear chicken broth
1/2 tsp corn starch

METHOD

1. Put rinsed fish maw in 1 litre of water with ginger juice, bring to a boil, cover, turn off heat, steeping for 30 minutes, rinse with cold water, squeeze dry and cut into about 5 cm sections.

2. Shred dried scallops.

3. Soften dried shrimps in cold water. Soften dried black mushrooms in cold water, remove stems, and cut mushrooms in halves. Cut carrot into thin slices. Peel ridges of the luffa skin, rinse and roll cut into small chunks.

4. Peel and brown garlic cloves in 1 tbsp of oil over low heat, add fish maw, dried shrimps, mushrooms, carrot, soy sauce, sugar and chicken broth. Braise for about 15 minutes, thicken sauce with corn starch and top with shredded scallops.

Fish Maw and Corn Soup

Serving: for 4   Preparation time: 30 minutes   Cooking time: 10 minutes

*Chinese version refer to p.94

INGREDIENTS

30 g dried scallops
100 g dried fish maw
1 can creamy style corn (418 g)
150 g minced pork
1 tbsp ginger juice
250 ml clear chicken broth
1 tsp salt
1/2 tsp corn starch
1/2 tsp white pepper

METHOD

1. Rinse fish maw and add to 1 litre of water with ginger juice, bring to a boil, cover and turn off the heat. Allow to simmer in residual heat for 30 minutes.

2. Rinse fish maw with cold water, squeeze out excess water and cut into bite size.

3. Prepare shredded scallops.

4. Mix minced pork with 1/2 tsp of salt and corn starch.

5. Boil chicken broth in a pot, add minced pork, shredded scallops, creamy corn and 125 ml of water, re-boil, and separate minced pork with chopsticks to prevent forming into lumps.

6. Add fish maw and cook for about 3 minutes. Flavor with 1/2 tsp of salt and white pepper. Ready to serve.

Double Boiled Chicken with Dried Fish Maw and Conch

Chinese version refer to p.98

INGREDIENTS
60 g dried fish maw, 250 g conch meat (frozen), 1 tsp salt, 1 tsp corn starch
1/2 chicken (about 600 g), 150 g pork shin, 20 g ginger slices
1 tsp Shaoxing wine, boiling water, salt for flavoring

METHOD
1. Prepare dried fish maw, and cut into pieces.

2. Defrost conch meat and clean with salt and corn starch, cut into slices, and blanch for 1 minute. Rinse and drain.

3. Blanch chicken for 1 minute, rinse and drain.

4. Cut pork into 2 cm pieces, and blanch for 1 minute. Rinse and drain.

5. Put chicken, pork, conch meat, ginger and wine into a stew container, add boiling water to cover ingredients, seal and double boil or steam over high heat for 2 hours.

6. Soften fish maw in hot water and add to the stew container, and continue to cook for 15 minutes. Flavor with salt.

 # Sweet Soup with Fish Maw and Dried Bean Curd Sheets

Chinese version refer to p.100

INGREDIENTS
160 g fish maw (pre-soaked), 75 g dried bean curd sheets (broken)
30 g raw coix seeds, 20 shelled gingko nuts, 60 g rock sugar, 500 ml boiling water

METHOD
1. Rinse coix seeds, add 500 ml water and 10 g rock sugar, bring to a boil, reduce to low heat and cook for 1 hour. Drain.

2. Rinse gingko nuts, separate each into two halves and remove the core. Rinse, add 500 ml water and 10 g rock sugar, bring to a boil, reduce to low heat and cook for 30 minutes. Drain. Cut fish maw into small pieces, blanch and drain.

3. Add broken bean curd sheets and the remaining rock sugar into 500 ml boiling water, stir, and cook over medium heat for 15 minutes. Put in fish maw and cook another 15 minutes until milky white, add gingko nuts and coix seeds and re-boil.

Fish Maw

Steamed Dried Scallops with Black Moss

Serving: for 6 | Preparation time: 30 minutes | Cooking time: 1 hour

*Chinese version refer to p.106

INGREDIENTS
50 g dried scallops, 30 g black moss, 200 g lettuce, 12 cloves garlic
1 tsp ginger juice, 1 tsp Shaoxing wine, 1 tbsp top soy sauce, 1 tsp sugar
250 ml clear chicken broth, 1 tbsp oyster sauce, 1/2 tsp corn starch, 2 tsp sesame oil

METHOD
1. Soak black moss in warm water for 30 minutes with a few drops of oil added. Rinse in fresh water and squeeze dry.

2. Add soy sauce, sugar and 1 tbsp of oil to 100 ml of chicken broth, stir until sugar has dissolved, and mix with black moss.

3. Peel and brown garlic in oil.

4. Rinse and put scallops into a 14 to 15 cm diameter bowl, add water to cover the scallops and soak for 5 minutes. Pour out water, and tear out and discard the small piece of tough adductor muscle. It will come off easily by simply pushing it downward along the side of the scallop.

5. Surround scallops with garlic, add ginger juice and wine, and fill with chicken broth to cover the scallops.

6. Place black moss on top and press firmly. Steam over high heat for 60 minutes.

7. Break up lettuce and blanch.

8. Turn the bowl over at the center of a large plate to transfer the scallops and black moss. Line the side of the plate with lettuce.

9. Boil remaining chicken broth, add oyster sauce and thicken with corn starch. Add sesame oil and dribble sauce on top of scallops, black moss and lettuce.

Shark's Fin with Dried Scallops and Eggs

Serving: for 4   Preparation time: 1 hour 15 minutes   Cooking time: 15 minutes

*Chinese version refer to p.108

INGREDIENTS
40 g dried scallops
120 g shark's fins, pre-soaked
125 ml clear chicken broth
75 g bean sprouts
6 egg yolks
1/2 tsp salt

METHOD
1. Soak scallops for 1 hour, drain, tear into shreds, and remove excess moisture with kitchen towels. Reserve half of the shredded scallops and deep fry the remainder.

2. Braise shark's fins with chicken broth and 1/4 tsp of salt. Drain.

3. Blanch bean sprouts for 5 seconds and rinse immediately with cold water.

4. Add 1/4 tsp salt to egg yolks and beat thoroughly. Heat 1 tbsp of oil to medium low temperature in a wok, change to low heat, put in egg yolks, use the bottom of a large metal ladle and press the eggs using a circular motion until the eggs are broken into very small bits. Add shark's fins and the reserved scallops, stir fry, put in bean sprouts and toss to mix thoroughly. Transfer to plate.

5. Sprinkle crispy shredded scallops.

Stuffed Hairy Gourd with Dried Scallops

Serving: for 6   Preparation time: 1 hour   Cooking time: 20 minutes

*Chinese version refer to p.111

INGREDIENTS

6 large dried scallops
2 hairy gourds
125 ml clear chicken broth
1 tbsp oyster sauce
1 tsp corn starch
1 tsp sesame oil

METHOD

1. Soak dried scallops in water for 30 minutes and steam another 30 minutes. Remove scallops and reserve the scallop water.

2. Scrape the skin off the gourds, cut into 6 equal sections with each one about 1/2 cm taller than the scallops. Carve out a round hole in the gourd about the size of the scallops without cutting through the bottom of the gourd.

3. Insert one scallop each into the gourd, transfer to a deep plate and steam for 20 minutes. Pour out water from the plate.

4. Boil chicken broth, the reserved scallop water and oyster sauce, and thicken with corn starch. Stir in sesame oil and pour over the stuffed gourd.

Steamed Minced Pork Pie with Salted Fish

Serving: for 4   Preparation time: 90 minutes   Cooking time: 10 minutes

*Chinese version refer to p.114

INGREDIENTS

75 g salted fish
30 g dried scallops
250 g minced lean pork
100 g pork fatback
1 tbsp corn starch
2 tbsp oat meal
1/2 tsp top soy sauce
1/2 tsp salt
1 tsp sugar
1 tbsp oil
1/2 tsp sesame oil
5 g shredded ginger

METHOD

1. Rinse and wrap pork fatback in cling wrap, and lay flat in the freezer for 60 minutes. Cut into small bits and chop briefly.

2. Chop minced lean pork further to obtain a finer texture.

3. Soak dried scallops in 5 tbsp of water (75 ml) until soft, and tear into shreds. Filter water for later use.

4. Marinate minced lean pork in a large bowl with soy sauce, salt, sugar and water from soaking scallops for 15 minutes.

5. Stir minced pork with chopsticks along a single direction until it becomes a gluey patty, pick up patty and smash it against the bowl 10 times or more.

6. Add crushed oat meal and fatty pork into minced pork and mix well.

7. Stir in scallops and corn starch, and mix in oil and sesame oil.

8. Place minced pork into a deep plate, smooth out the surface with wet fingers, seal tightly with microwave oven wrap and steam over high heat for about 10 minutes.

9. Pan fry salted fish over low heat in a small amount of oil, and put on top of meat pie, and top with shredded ginger.

Dried Scallops

Dried Scallops and Chicken Soup

Serving: for 4 | Soaking time: 2 hours | Preparation time: 15 minutes | Cooking time: 10 minutes

*Chinese version refer to p.118

INGREDIENTS
30 g dried scallops
150 g chicken breast
1 egg white
500 ml chicken broth
1/2 tsp salt
1/2 tsp white pepper
2 tbsp water chestnut starch
1 tsp sesame oil

METHOD
1. Soak dried scallops for 2 hours and tear into shreds.
2. Remove any tendons and membranes from the chicken breast and chop chicken into a very fine paste. Put into a large bowl.
3. Stir egg white into the chicken paste, and remove any tendon or membrane found.
4. Boil 250 ml chicken broth, pour into the bowl containing the chicken paste and stir in a single direction until the chicken paste is fully blended with the broth to become a chicken paste broth.
5. Boil the remaining chicken broth, add shredded scallops and cook for about 3 minutes. Season with salt and pepper, make a wet starch with water chestnut starch and stir gradually into the soup to thicken.
6. Reduce to low heat, add chicken paste broth gradually, stir gently and turn off the heat. Add sesame oil and put into a large soup serving bowl.

Steamed Mantis Shrimps and Pork with Shrimp Paste

Serving: for 4　Preparation time: 10 minutes　Cooking time: 10 minutes

*Chinese version refer to p.122

INGREDIENTS

25 g dried mantis shrimps, 150 g pork shoulder, 1/2 tsp shrimp paste
1 tsp sugar, 1/2 tbsp ginger juice, 1/2 tbsp Shaoxing wine
1 tsp corn starch, 5 g shredded ginger

METHOD

1. Rinse and soak dried mantis shrimps for 10 minutes, drain.

2. Slice pork, mix with shrimp paste, sugar, ginger juice, wine and corn starch, and transfer to plate. Arrange mantis shrimps on the pork top with shredded ginger, and steam for 10 minutes. Drizzle 1 tbsp heated oil on top.

 # Rice with Dried Shrimps, Dried Squids and Pork in a Casserole

Serving: for 4　Preparation time: 30 minutes　Cooking time: 30 minutes

*Chinese version refer to p.124

INGREDIENTS

320 g rice, 45 g dried shrimps, 30 g small semi-dried squids
150 g pork shoulder, 1/4 tsp salt, 1/4 tsp sugar
1/2 tsp top soy sauce, 1/2 tsp corn starch, 1 tbsp chopped garlic

METHOD

1 Rinse rice and drain.

2 Soak dried shrimps for 30 minutes, drain.

3 Soak semi-dried squids for 15 minutes, drain.

4 Slice pork, and marinate with salt, sugar, soy sauce, corn starch and 2 tbsp of water for 15 minutes.

5 Heat 2 tbsp of oil in a casserole, stir fry garlic until pungent, and add rice and stir fry together for about 1 minute. Add water to about 2 cm above the rice, bring to a boil, open the cover and put in dried shrimps, semi-dried squids and pork, and boil until most of the water has evaporated. Cover, reduce to low heat and continue to cook for about 15 minutes.

Dried Shrimps and Dried Mantis Shrimps

Stewed Sea Cucumber with Dried Cuttlefish and Pork

Serving: for 4 | Soaking time: 2 hours | Preparation time: 10 minutes | Cooking time: 2 hours

*Chinese version refer to p.134

INGREDIENTS
150 g dried cuttlefish
600 g pork belly
250 g sea cucumber, pre-soaked
50 g ginger slices
1 tbsp bean sauce
125 ml Shaoxing wine
2 tbsp top soy sauce
1 tbsp sugar
1 tsp salt
1 tsp corn starch

METHOD
1. Soak dried cuttlefish for 15 minutes, remove cartilage, soak for another 10 minutes and cut into squares.
2. Blanch pork belly and cut into 2 x 2 cm pieces.
3. Blanch sea cucumber and cut into 2 cm pieces.
4. Heat 2 tbsp of oil, stir fry ginger and bean sauce, add cuttlefish and put in pork. Sprinkle wine and add water to cover all ingredients, bring to a boil, reduce to low heat and cook for 1.5 hours.
5. Add sea cucumber, soy sauce, salt and sugar, braise for 15 minutes and reduce the sauce over high heat. Thicken sauce further with corn starch.

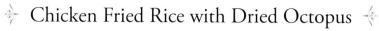

Chicken Fried Rice with Dried Octopus

Serving: for 4   Preparation time: 20 minutes   Cooking time: 5 minutes

*Chinese version refer to p.137

INGREDIENTS

3 bowls cooked rice
100 g dried octopus
200 g boneless chicken thigh
1 tbsp chopped ginger
1 stalk spring onion, chopped
1 tbsp top soy sauce

MARINADE FOR CHICKEN

1 tsp top soy sauce
1/2 tsp sugar
1 tsp corn starch

METHOD

1. Rinse rice with cold water to wash away some of the starch, break up lumps and put into a colander.

2. Cover dried octopus with water and steam for 15 minutes, cut into small bits and mix with 1 tsp of oil.

3. Dice chicken and marinate for 15 minutes.

4. Stir fry ginger until pungent in 2 tbsp of oil, add chicken and octopus, put in rice and soy sauce and stir fry. Finally add spring onion and toss thoroughly.

Dried Octopus Soup with Lotus Roots and Pork

Serving: for 4 | Preparation time: 15 minutes | Cooking time: 2 hours 45 minutes

*Chinese version refer to p.140

INGREDIENTS

60 g dried octopus, 600 g pork knuckle, 600 g lotus roots
50 g mung beans, 15 g ginger slices, 2.5 litres water, 1 tsp salt

METHOD

1. Rinse and soak dried octopus in warm water for about 1 hour.

2. Rinse and blanch pork, drain.

3. Separate lotus roots at the knots, and cut off the ends at about 1 cm. Clean the channels inside the lotus roots.

4. Rinse and soak mung beans for 30 minutes, and stuff lotus roots channels with mung beans using a chopstick until about 80% full.

5. Boil about 2.5 litres of water in a large pot, put in ginger, pork, lotus roots and dried octopus, cover and boil for 10 minutes over high heat. Skim froth and scum from the surface, reduce to low heat and cook for 2.5 hours. Add salt to flavor.

Steamed Dried Squids with Ginger Juice

Serving: for 4   Preparation time: 30 minutes   Cooking time: 5 minutes

*Chinese version refer to p.142

INGREDIENTS
100 g dried squids
1 tbsp ginger juice
1 tsp Shaoxing wine
1 tsp sugar
1 tbsp oil
10 g shredded ginger
1 tsp top soy sauce
1 tbsp chopped spring onion

METHOD
1. Rinse squids, and soak for 30 minutes. Drain.

2. Mix in ginger juice, wine, sugar and 1 tbsp oil, transfer to a plate, top with shredded ginger and steam for 5 minutes. Drizzle soy sauce on top.

Steamed Pork Pie with Dried Squid

Serving: for 4 | Preparation time: 2 hours 15 minutes | Cooking time: 10 minutes

Chinese version refer to p.144

INGREDIENTS
50 g dried squid
2 water chestnuts
75 g pork fatback
200 g minced lean pork
1/2 tsp salt
1/2 tsp sugar
1/2 tbsp top soy sauce
1/2 tbsp chopped garlic
1 tbsp corn starch
1/2 tsp white pepper
1 tbsp oat meal
1 tsp sesame oil

METHOD
1. Soak dried squid in warm water for 2 hours, rinse thoroughly, remove membrane and bone, and cut into small bits. Reserve the water.
2. Peel, wash and dice water chestnuts, and soak in cold water until use.
3. Wash pork fatback, and place into the freezer until firm. Cut into bits.
4. Chop minced lean pork further to obtain a finer texture.
5. Add 2 tbsp of water to the minced pork, mix in salt, sugar and soy sauce. Rest pork for 10 minutes and stir with chopsticks along a single direction until a gluey texture is reached.
6. Mix in squid, fatty pork, water chestnuts and garlic, and add corn starch and white pepper.
7. Add crushed oat meal, and mix in sesame oil. Form meat into a patty and place into a plate.
8. Steam over high heat for 10 minutes.

Dried Squid, Cuttlefish and Octopus

Stir-fried Squid with Hotbed Chinese Chives

Serving: for 4   Preparation time: 40 minutes   Cooking time: 5 minutes

*Chinese version refer to p.146

INGREDIENTS
1-2 fresh squid (about 500 g)
1 semi-dried squid
1 tbsp ginger juice
150 g hotbed Chinese chives, cut into 5 cm sections
4 cloves garlic, chopped
2 shallots (quartered), 1 tsp salt
1/2 tsp sugar, 1 tsp Shaoxing wine
1/4 tsp sesame oil, 1 bowl ice cold water
1/2 tsp corn starch

METHOD
1. Cut open the fresh squid from the stomach side and remove all viscera, tear off and discard the fins and the membrane on the outside. Cut squid down the centre into two pieces.

2. Place one piece of squid stomach side up on a piece of kitchen paper towel on top of the cutting board, hold the squid flat with one hand and make shallow crisscross cuts with a knife in the other hand. Cut squid crosswise into 6 cm pieces. Repeat the above with the other piece of squid.

3. Soak semi-dried squid in water for 30 minutes, change water once, rinse thoroughly, and prepare using the same method as for fresh squid.

4. Boil water in a wok, add ginger juice, put semi-dried squid into the water and turn off the heat immediately. Disperse squid pieces in the water with chopsticks and take out after about 3 seconds. Immediately immerse squid in a bowl of ice-cold water. Drain when the water is no longer cold.

5. Re-boil water in a wok, put in fresh squid, and turn off the heat immediately. Disperse squid pieces in the water with chopsticks and take out after about 5 seconds. Immediately immerse squid in a bowl of ice-cold water. Drain when the water is no longer cold.

6. Heat 2 tbsp of oil to medium high temperature over high heat, stir fry shallots and garlic until pungent, put in squids and salt, toss a few times, add wine, sugar and hotbed Chinese chives, and toss rapidly. Finally stir in sesame oil and thicken lightly with corn starch.

Dried Squid, Cuttlefish and Octopus

Stir Fried Dried Squid, Dried Prawns with Flowering Chinese Chives

*Chinese version refer to p.149

INGREDIENTS
50 g semi-dried squid, 30 g dried tiger prawns, 150 g pork belly,
200 g flowering Chinese chives
1 red sweet pepper, 1 tsp satay sauce, 1 tbsp Shaoxing wine
1 tsp top soy sauce, 1/8 tsp white pepper

METHOD
1. Soak dried squid for 30 minutes, remove membrane, and cut into thin strips along the width of the squid.

2. Soak dried tiger prawns for 20 minutes, and steam together with water for 10 minutes. Drain.

3. Blanch pork belly and cut into thin slices.

4. Rinse flowering Chinese chives and cut into 6 cm lengths. Deseed red pepper and cut into strips.

5. Heat 1 tbsp of oil and stir fry squid, dried prawns, pork belly and red pepper, add satay sauce and drizzle wine along the inside of the wok. Put in flowering Chinese chives and stir fry over high heat until most of the liquid has evaporated. Add soy sauce and white pepper and toss thoroughly.

度量衡換算表

　　市面上的食譜書，包括我們陳家廚坊系列，食譜中的計量單位，都是採用公制，即重量以克來表示，長度以厘米 cm 來表示，而容量單位以毫升 ml 來表示。世界上大多數國家都採用公制，但亦有少數國家如美國，至今仍使用英制（安士、磅、英吋、英呎）。

　　香港和澳門，一般街市仍沿用司馬秤（斤、兩），在香港超市則有時用公制，有時會用美制，所以香港是世界上計量單位最混亂的城市，很容易會產生誤會。與香港關係緊密的中國大陸，他們的大超市有採用公制，但一般市民用的是市制斤兩，這個斤與兩，實際重量又與香港人用的司馬秤不同。

　　鑒於換算之不方便，曾有讀者要求我們在食譜中寫上公制及司馬秤兩種單位，但由於編輯排版困難，實在難以做到。考慮到實際情況的需要，我們覺得有必要把度量衡的換算，以圖表方式來說清楚。

重量換算速查表 （公制換其他重量單位）

克	司馬兩	司馬斤	安士	磅	市斤
1	0.027	0.002	0.035	0.002	0.002
2	0.053	0.003	0.071	0.004	0.004
3	0.080	0.005	0.106	0.007	0.006
4	0.107	0.007	0.141	0.009	0.008
5	0.133	0.008	0.176	0.011	0.010
10	0.267	0.017	0.353	0.022	0.020
15	0.400	0.025	0.529	0.033	0.030
20	0.533	0.033	0.705	0.044	0.040
25	0.667	0.042	0.882	0.055	0.050
30	0.800	0.050	1.058	0.066	0.060
40	1.067	0.067	1.411	0.088	0.080
50	1.334	0.084	1.764	0.111	0.100
60	1.600	0.100	2.116	0.133	0.120
70	1.867	0.117	2.469	0.155	0.140
80	2.134	0.134	2.822	0.177	0.160
90	2.400	0.150	3.174	0.199	0.180
100	2.67	0.17	3.53	0.22	0.20
150	4.00	0.25	5.29	0.33	0.30
200	5.33	0.33	7.05	0.44	0.40
250	6.67	0.42	8.82	0.55	0.50
300	8.00	0.50	10.58	0.66	0.60
350	9.33	0.58	12.34	0.77	0.70
400	10.67	0.67	14.11	0.88	0.80
450	12.00	0.75	15.87	0.99	0.90
500	13.34	0.84	17.64	1.11	1.00
600	16.00	1.00	21.16	1.33	1.20
700	18.67	1.17	24.69	1.55	1.40
800	21.34	1.34	28.22	1.77	1.60
900	24.00	1.50	31.74	1.99	1.80
1000	26.67	1.67	35.27	2.21	2.00

司馬秤換公制

司馬兩	司馬斤	克
1		37.5
2		75
3		112.5
4	0.25	150
5		187.5
6		225
7		262.5
8	0.5	300
9		337.5
10		375
11		412.5
12	0.75	450
13		487.5
14		525
15		562.5
16	1	600
24	1.5	900
32	2	1200
40	2.5	1500
48	3	1800
56	3.5	2100
64	4	2400
80	5	3000

英制換公制

安士	磅	克
1		28.5
2		57
3		85
4	0.25	113.5
5		142
6		170
7		199
8	0.5	227
9		255
10		284
11		312
12	0.75	340.5
13		369
14		397
15		426
16	1	454
24	1.5	681
32	2	908
40	2.5	1135
48	3	1362
56	3.5	1589
64	4	1816
80	5	2270

容量

量杯	公制（毫升）	美制（液體安士）
1/4 杯	60 ml	2 fl. oz.
1/2 杯	125 ml	4 fl. oz.
1 杯	250 ml	8 fl. oz.
1 1/2 杯	375 ml	12 fl. oz.
2 杯	500 ml	16 fl. oz.
4 杯	1000 ml /1 公升	32 fl. oz.

量匙	公制（毫升）
1/8 茶匙	0.5 ml
1/4 茶匙	1 ml
1/2 茶匙	2 ml
3/4 茶匙	4 ml
1 茶匙	5 ml
1 湯匙	15 ml

作者簡介
陳紀臨、方曉嵐

　　陳紀臨、方曉嵐夫婦，是香港著名食譜書作家、食評家、烹飪導師、報章飲食專欄作家。他們是近代著名飲食文化作家陳夢因（特級校對）的兒媳，傳承陳家兩代的烹飪知識，對飲食文化作不懈的探討研究，作品內容豐富實用，文筆流麗，深受讀者歡迎，至今已在香港出版了十多本食譜書，作品遠銷海外及國內市場，更在台灣多次出版。

　　2016 年陳紀臨、方曉嵐夫婦應出版商 **Phaidon Press** 的邀請，用英文為國際食譜系列撰寫了 *China The Cookbook*，介紹全國 33 個省市自治區的飲食文化和超過 650 個各省地道菜式的食譜，這本書得到國際上好評，並為世界各大主要圖書館收藏。這本書的中文、法文、德文、西班牙文版現已出版，將會陸續出版意大利文、荷蘭文等，為中國菜在國際舞台上作出有影響力的貢獻。

林長治

　　出生於魚翅世家，香港祐生魚翅公司創辦人，對海味乾貨具豐富的知識，特別是對海參的研究。

　　曾兼任香港餐務管理協會會董，美國駐香港環球國際企業顧問（1989-2009）。

鳴謝

LEGLE
PORCELAIN

https://leglefrance.com/

R U Y I

Eastern Philosophy
Modern Sensibility

https://ruyi.global/

著者	**Author**
方曉嵐、陳紀臨、林長治	Diora Fong, Keilum Chan, George Lam
責任編輯	**Editor**
譚麗琴	Catherine Tam
攝影	**Photographer**
梁細權	Leung Sai Kuen
裝幀設計	**Design**
羅美齡	Amelia Loh
排版	**Typesetting**
辛紅梅	Cindy Xin
出版者	**Publisher**
萬里機構出版有限公司	Wan Li Book Company Limited
香港北角英皇道499號	20/F, North Point Industrial Building,
北角工業大廈20樓	499 King's Road, Hong Kong
電話	Tel: 2564 7511
傳真	Fax: 2565 5539
電郵	Email: info@wanlibk.com
網址	Web Site: http://www.wanlibk.com
	http://www.facebook.com/wanlibk
發行者	**Distributor**
香港聯合書刊物流有限公司	SUP Publishing Logistics Hong Kong Limited
香港荃灣德士古道220-248號	16/F., Tsuen Wan Industrial Centre, 220-248 Texaco Road,
荃灣工業中心16樓	Tsuen Wan, N.T., Hong Kong
電話	Tel: 2150 2100
傳真	Fax: 2407 3062
電郵	Email: info@suplogistics.com.hk
承印者	**Printer**
中華商務彩色印刷有限公司	C & C Offset Printing Co., Ltd.
香港新界大埔汀麗路36號	
出版日期	**Publishing Date**
二〇二一年七月第一次印刷	First print in July 2021

陳家廚坊
Chan's Kitchen

陳家廚坊
Chan's Kitchen